ISWA 88
PROCEEDINGS
of the
5th International Solid Wastes Conference
International Solid Wastes

ISWA 88
PROCEEDINGS
of the
5th International
Solid Wastes Conference

International Solid Wastes
and Public Cleansing Association

September 11–16th, 1988
Copenhagen, Denmark

VOLUME 2
POSTER PRESENTATIONS

Edited by
Lizzi Andersen and Jeanne Møller
Vester Farimagsgade 29
DK 1606 Copenhagen V
Denmark

1988

ACADEMIC PRESS

Harcourt Brace Jovanovich, Publishers
London   San Diego   New York   Berkeley
Boston   Sydney   Tokyo   Toronto

ACADEMIC PRESS LIMITED
24/28 Oval Road,
London NW1 7DX

United States Edition published by
ACADEMIC PRESS INC.
San Diego, CA 92101

Copyright © 1988, by
ACADEMIC PRESS LIMITED

All rights Reserved
No part of this book may be reproduced in any form by photostat, microfilm,
or any other means, without written permission from the publishers except
pages 137–144, 261–266, 287–294 where copyright is held by UKAEA
and pages 393–408 where copyright is held by ISWA.

ISBN 0-12-058452-2

Printed in Great Britain by St Edmundsbury Press Limited,
Bury St Edmunds, Suffolk

PREFACE

The ISWA Quadrennial Conference and Exhibition was established to provide an international forum for current topics in the field of solid wastes management, technology and research.

With its wide range of topics, the ISWA 88 Conference offers unique facilities for the presentation and exchange of novel technology and thoughts both in a wide context and on more specific aspects.

These Proceedings contain papers selected from 254 submitted abstracts by the programme committee. The ISWA 88 General Secretariat acknowledges the 140 authors for their cooperation and timely submittal of camera-ready manuscripts.

Lizzi Andersen
Secretary General

Jeanne Møller
Administrative Secretary

ISWA 88 General Secretariat
Vester Farimagsgade 29
DK 1606 Copenhagen V, Denmark

ISWA 88

Thank you for registering for the ISWA 88 Congress. We would like to welcome you and we hope that the content of the conference programme and the leisure activities will match your expectations.

Apart from the speeches by representatives from the World Bank, the World Health Organization, and the United Nations Environmental Programme, you will also hear contributions from specialists from 28 countries. They will deal with the most far-reaching areas of concern within ISWA's domain. I sincerely hope that each contribution will be followed by discussions with frank exchanges of views.

Although the contributors only have a limited time to speak, each one has put in a great deal of time and effort in the preparations. We would like to thank all the speakers for their participation in the success of this major ISWA event.

J. DEFÈCHE
President of ISWA

Contents

Session 1	**ISWA 88 Opening Session**	1
Session 2	**Public Attitudes Towards Solid Waste Management**	3
Session 3	**Biological Treatment of Solid Waste**	5

F. Cayrol, C. Claquin and J-P. Peillex, France: Anaerobic Digestion of Municipal Solid Waste by the Valorga Process. ... 7

J. L. Dorronsoro, R. Muñoz, A. Sanchez, M. L. Solano and B. Torralba, Spain: Microbiological Treatment of Liquid and Solid Wastes of Urban Source. ... 13

I. W. Koster, D. A. Vroon and J. Bovendeur, The Netherlands: Earthworm-Filters for Treatment of Liquid Sludge from Fishfarms. ... 19

H-H. Spendlin and R. Stegmann, FRG: Anaerobic Fermentation of the Vegetable, Fruit, and Yard Waste. ... 25

Session 4	**Sanitary Landfills**	33

T. Assmuth and M. Melanen, Finland: Screening Toxicants in Waste Deposit Runoff. ... 35

E. Biener and T. Sasse, FRG: Experience with New Techniques in Slurry Cutoff Wall Construction. ... 41

G. Bergvall, Sweden: DEPÅ 90 – Progress Report from a Swedish R & D Landfill Project. ... 47

G. Bergvall, R. Karlsson and S. Wallin, Sweden: Measurement of Mercury Vapor Emissions from Swedish Waste Landfills. ... 55

T. Bramryd, Sweden: Leachate from Landfills – A Valuable Fertilizer for Revegetation of Landfills with Grass or Energy Forests. ... 61

H. Doedens and B. Weber, FRG: Degasification of a Capped Landfill Site for Domestic Refuse. ... 69

P. Fletcher, UK: Design and Early Results from the Landfill Gas Enhancement Test Cells at Brogborough, Bedfordshire, United Kingdom. ... 75

O. Holmstrand and B. Troedsson, Sweden: Leachate Purification by Irrigating a Peat Bog. 81

L. M. Johannesen, Denmark: Disposal of Gypsum from Flue Gas Desulfusization (FGD) in Land Reclamation. 87

B. J. W. Manley and H. S. Tillotson, UK: Improving the Effectiveness and Reliability of Landfill Gas Pumping Trials. 93

A. Mennerich, K. Kruse, O. Schüttemeyer and R. Damiecki, FRG: Enhanced Leaching and Decomposition of a Sanitary Landfill to Minimize its Future Pollution Potential. 99

E. Parráková, I. Fratrič and J. Mayer, CSSR: Long-Term Disposal Facilities for Chemical Wastes with Respect to Environmental Protection. 107

M. Schäfer and H. J. Schröter, FRG: Adsorptive Removal and Recovery of Halocarbons from Landfill Gas. 113

J. Schneider, FRG: Landfill Gas – Environment Protection and Energy Recovery by a 13.5 MW Total-Energy-Station. 119

P. E. Scott, C. G. Dent and G. Baldwin, UK: A Study of Trace Components in Landfill Gas from Three UK Household Waste Landfill Sites. 125

L. Sridharan and P. Didier, USA: Leachate Quality from Containment Landfills in Wisconsin. 133

P. P. Stecker, USA: Advanced Landfill Technology for Joint Leachate and Gas Management. 139

M. Wahlström and J. Ranta, Finland: Leaching of Metals from Desulphurization Wastes. 145

M. Werner and G. Olderdissen, FRG: Removal of Chlorinated Hydrocarbons from Landfill Gas. 151

Chr. Wolffson, FRG: Laboratory Scale Test for Anaerobic Degradation of Municipal Solid Waste. 159

Session 5   Sludge Disposal and Landuse 167

M. Baldi, G. Chierico, V. Riganti and M. Specchiarello, Italy: Laboratory Experiences on the Formation of Percolate from Civil and Industrial Conditioning Sludges. 169

M. Fritze, Denmark: Treatment of Sludge in an LGW Plant, System Vølund. 177

D. Holtz, FRG: Production of an Earthlike Material by Combining Sewage Sludge with Porous Mineral Substances. 187

E. S. Kempa and T. Marcinkowski, Poland: Liming of Sludge from Biological Treatment of Cokery Wastewater. 193

T. Marcinkowski, Poland: Application of Sludge-Lime Mixtures to Soil Reclamation: Vegetation Tests. 199

M. Ottaviani, L. Olori and G. Basile, Italy: Heavy Metals in Sewage Sludge: Concentration Variability and Elution Possibility. 207

K. Valdmaa, Sweden: Long-Term Field Trials with Sewage Sludge in Comparison with Farmyard Manure and Commercial Fertilizer. 213

Session 6   Hazardous Waste 221

M. Bailey, New Zealand: New Zealand's Hazardous Waste Management Strategy. 223

K. Christiansen and E. Hansen, Denmark: Substitution of Hazardous Constituents of Household Products. 229

F. Crotogino and H. J. Schneider, FRG: Safe, Zero-Immission Ultimate Disposal of Solid Hazardous Wastes in Salt Caverns. 237

M. Ivanc and B. Pleskovič, Yugoslavia: Case Study of the 250.000 m^3 Acid Tar Disposal Pit above the Rijeka Port. 245

C. Journault and J-P. Plamondon, Canada: Managing Sites Contaminated by Hazardous Waste in Quebec. 251

K. Kawamura, Japan: Involvement of Municipal Government in Appropriate or Environmentally Acceptable Treatment and Disposal of Hazardous or Toxic Industrial Wastes in Osaka. 259

R. Louw, J. A. Manion and J-P. Born, The Netherlands: Coal Mediated vs. Thermal Hydrogenolysis of Chlorinated Wastes. 267

G. Mininni, V. Lotito, M. Santori and L. Spinosa, Italy: Hazardous Sludge Management in Italy. 273

T. Morishita, Japan: Hazardous Wastes Management in Japan. 279

W. Smykatz-Kloss and L. Kaeding, FRG: Interactions Between Clay Minerals and Organic Compounds Around Hazardous Waste Landfills. 283

Session 7   Computerized Solid Waste Management 289

J. Kaila, Finland: Mathematical Model for Strategy Evaluation of
Waste Management Systems. 291

A. Peroni, A. Magagni and P. Predieri, Italy: Computerizing in the
Management of Urban Cleansing Municipalities Services and
Plants in Italy. 299

Session 8   Thermal Treatment of Solid Waste 305

R. J. Alvarez, USA: Update on Generation of Energy from MSW in
the United States. 307

J. R. Barton and D. Scott, UK: Pretreated Refuse Incineration,
Impact on Energy Recovery, Costs and Emissions. 313

H. Braun and R. Seitner, FRG: Mercury Analyzer for Continuous
Surveillance of the Waste Water of Refuse Incineration Plants. 319

A. Berbenni, C. Collivignarelli, F. Nobile, A. Bassetti and
A. Farneti, Italy: Detoxification of Dusts Derived from Urban
Solid Wastes Incineration Plants. 325

C. Dimitrov and D. Baltzinger, France: The New Saint Ouen's
Incineration Plant (France). 335

P. Kahanek, A. Merz and K. Schuy, FRG: An Isokinetic Sampling
System for Continuous Long-Term Monitoring of Hazardous Flue
Gas Emissions. 339

K. Keldenich, H. Ruppert and R. Jockers, FRG: Mercury Removal
from Waste Gases. 345

J. P. Léglise, D. Abraham and D. Marchand, France: A New
Technique for Reducing Hydrochloric Acid and Heavy Metal
Emissions Resulting from Household Waste Incineration. 353

P. Mostbauer, Austria: Hazardous Scrubber Waste Disposal at
Vienna. 359

T. Obermeier and L. Freutsmiedl, FRG: Experiences with Chlorine
Balancing at an RDF Plant. 369

P. Souet, France: Energy Valorization of Used Tyres. 377

Session 9   Solid Waste Management in Developing Countries 385

M. T. Gadi, The Philippines: Solid Waste Management in Metro
Manila: A Systems Approach. 387

Session 10   Collection and Public Cleansing 393

P. Masselink, The Netherlands: Introduction of Recycling Based Separated Collection Programs in Relation to Waste Disposal Methods. 395

Session 11   Resource Recovery and Recycling 401

P. E. O. Berg, Sweden: Recyclables in Municipal Waste – manuscript not received in time for publication.

P. E. O. Berg, Sweden: Municipal Waste Reduction by Recycling in Sweden (Today and Tomorrow) – manuscript not received in time for publication.

A. Elmlund, Denmark: Waste Reduction Through Recycling in Denmark Year 2000. 403

A. Reichert and H. Hoberg, FRG: Electro-Optical Sorter for Waste Glass. 413

Session 12   Waste Reduction and Clean Technology 419

J. Kappel, A. Kaiser and D. Sporn, FRG: Investigations concerning the Direct Remelting of Filter Dusts from Glass Industry. 421

J. Kryger, Denmark: Information System for Cleaner Technologies. 427

Session 13   International Activities in Solid Waste Management 433

Session 14   Local, Regional and National Solid Waste Management   435

T. Bresitsláv̂, CSSR: General Plan of Solid Wastes Treatment in Prague. 437

H. Juvonen and J. Kaila, Finland: Source Separation of Municipal Solid Wastes – Finnish Experience of Two Case Studies. 443

F. Koza and I. Krüger, Denmark: R–98 Plants for Recovery of Resources by Manual Sorting of Pre-Sorted Waste. 449

M. Lamblin, D. Verbecke, D. Lemarchand and J-M. Tayon, France: Creation of a Facility for the Fabrication of Substitute Fuel from Industrial Waste Liquids, Sludges and Sounds Produced by EEC Member Countries. 457

P. Lingwood, J. Boxall and B. Ashcroft, Hong-Kong: Conceptual Design and the Environment – A Hong Kong Example – manuscript not received in time for publication.

W. Lutz, Austria: Combined Energy Recovery and Recycling of
Wastes in Smaller Regions. 463

K. Niederl, Austria: Integrated Refuse Management from a
Community Viewpoint: The City of Graz, Austria. 471

E. Papachristou, Greece: Solid Wastes Management in Rhodos. 479

P. O. Scokart, M. P. Walbrecq and J. P. Bas, Belgium: An Example
of Sewage Sludge Utilization in Agriculture at the Municipal
Level in Belgium. 487

Session 15   ISWA 88 Closure 493

Session 1

ISWA 88 Opening Session

Session 2

Public Attitudes Towards Solid Waste Management

Session 3

Biological Treatment of Solid Waste

ANAEROBIC DIGESTION OF MUNICIPAL SOLID WASTE BY THE VALORGA PROCESS

CAYROL François, CLAQUIN Catherine and PEILLEX Jean-Paul

VALORGA SA, 5 rue de Massacan, BP 56, 34740 VENDARGUES, FRANCE.

Conventional processing methods for the elimination of urban waste each only consider one aspect of the deposit : waste for storage (placing in dumps), combustible waste (incineration) and enrichment (composting). Furthermore, their quantity and their quality varies according to country, city, district, season, etc.

The following table gives an indication, by way of example, of the "average" composition of this deposit in France (source : ANRED).

Composition	percentage	Composition	percentage
Paper/cardboard	20 to 35	Plastics	3 to 60
Fermentables	15 to 35	Textiles	1 to 6
Glass	5 to 10	Fine elements	10 to 20
Metals	5 to 8	Sundry	complement to 100

<u>A logical and flexible process.</u> The VALORGA process optimises the treatment of each fraction :
- Removal of heavy inert substances and metals prior to fermentation
- Transformation of the rapidly biodegradable matter into biogas
- Agronomic valorisation of the residual biodegradable organic matter

- Thermic valorisation of the organic fraction which is not essentially biodegradable (plastics, synthetic fibres, wood) (cf. Diagram n°1 - Principle).

It is thus possible to find an efficient solution to the majority of cases arising by simply adapting the size of the element processing units depending on the special composition of the urban waste for a site, on the local facilities for valorisation of the products and on the possible existence of separative collection methods.

<u>Based on a novel fermentor.</u> In the early '80s, numerous processes for continuous methanisation of deposits and liquid effluent existed; they were not able to exceed an approximate

limit of 10% of dry matter, and this prevented the processing of solid waste or that having a high content of dry matter (35 to 40 % in the case of domestic waste). The VALORGA fermentor is approved for processing matter in the form of a thick magma. Resorting to the use of biogas which is compressed and injected into the matter, disposes with the need for internal mechanical parts (cf. Diagram n°2). This thus effects the mixing process which is required for thermic homogeneity, for replacing the exchange surfaces and for piston-type development of the matter between entry and exit, which itself is facilitated by an internal wall (cf. Diagram n°3).

<u>Offering numerous advantages.</u> The high viscosity of this matter disposes with the need for separating the inert and heavy matter which is still present in the fermentation tank; the process is able to effect a simplified sorting operation beforehand. A reduction by a factor of 4 to 5 in the volume of the fermenting room in comparison to fermentors for liquid matter of the infinitely mixed type, makes it possible to greatly reduce not only the ground surface occupied and the investment costs, but also loss in heat and in volume : work involving a high content of dry matter avoids the need to heat water unnecessarily (this provides an enormous energy-saving method in comparison to all processes in a liquid medium). It also enables the processing of other waste, whether or not it is associated with domestic waste : concentrated urban deposits, abattoir waste, and waste from the agricultural and alimentary industries. Lastly, this design allows the construction of fermentors having a volume of several thousand cubic metres which can operate by a mesophilic system (35 - 40°C) or by a thermophilc system (55 - 60°C).

<u>And reaching industrial maturity.</u> Less than three years have elapsed between the formation of the Company (June 1981) and commencement of operations (April 1984) by this industrial unit situated in La Buisse, near Grenoble (Isère), which immediately proved to very successful. Its factory at Amiens (Somme) was constructed between December 1987 and June 1988, and it is now in the process of starting up. It will have a processing capacity of 110 000 tonnes per year for urban waste, with three fermentors of 2 400 m^3 each. This installation has been set up with financial backing from the European Economic Community. To date, seven other factories are in the course of being set up.

The installation at La Buisse. Fermentation is carried out by the mesophilic system (37°C) in a digester with a working capacity of 400 m^3; it is designed to process 8 000 tonnes of crushed and graded waste per year supplied from an already existing composting installation; this waste is diluted to contain between 35 and 40 % of dry matter prior to entry and extracted containing between 28 and 33 % for an applied turnaround time of 15 to 20 days. The following table shows the main biological results under these conditions :

Organic matter introduced	(kg/m^3/day)	11.2 to 11.8
Biogas produced per tonne of milled matter	(m^3/t)	145 to 160
Biogas produced per tonne of dry matter introduced	(m^3/t VSC)	500 to 560
Biogas produced per working volume of fermenting room	(m^3/m^3/day)	4.2 to 4.8
Methane produced	(% biogas)	59 to 61

Depending on the day of the week, ratios are maintained of 0.6 to 1.4 between daily volume output and the average volume output (cf. Diagram n°4). A degredation can also be observed of approximately 50 % of the organic matter during the fermentation process.

The biogas is delivered in this condition to a manufacturer in the neighbourhood, where it is burnt without purification. The biogas produced at Amiens is delivered after purification, for introduction into the natural gas distribution system of Gaz de France. An accurate energy assessment covering 2 months, March and April 1986, makes it possible to check the level of thermic requirements for the fermentor : 128 m^3/day of biogas consumed for 1 612 m^3/day produced, i.e. 7.9 % ! Over the year, the calculation increases in terms of self-consumption by 5.6 % of production.

Agronomic valorisation of the fermented matter. Organic matter for agronomic use, so-called refined digestate, is the end product which remains after the removal of the last inert substances, which are mainly organic (combustible waste). Refined digestate is present in the form of a product which is dry (60 to 65 % dry matter), crumbly, consisting of fine regular grains (10 mm), odourless, and it is slightly basic (pH = 8 to 8.5). These properties allow very easy handling (loading, transportation, measuring) and possible pouring using all existing agricultural tools; the following table shows the average composition of the dry matter :

Composition of the dry matter   %

Organic matter (VSC)	30 - 35
Carbon	10 to 12
Nitrogen	0.8 to 0.9
Phosphorous (P2O5)	0.3
Potassium (K2O)	1.3
Calcium (Ca)	5.3
Magnesium (Mg)	0.3

It will be noticed that 25 % of the nitrogen content is present in ammoniacal form (immediately available to plants), and that the C/N ratio lies between 12 and 15 (with a product almost exempt from plastics). Through the action of removing the metals, glass and fine inert substances at the first sorting operation, this reduces, at the outset, from 45 to 55 % of the heavy metal content (Hg, Cd, Zn, Pb, Cr and Cu) by thus avoiding their partial solubilisation. The final purification sorting operation removes matter (plastics and textiles) which is rich in immobilized heavy metals (particularly Cd and Cr). The digestate conforms to current standards governing levels of heavy metals for the majority of European countries. The novel solutions which are approaching the end of their development phase, will make it possible to guarantee in the near future a still large qualitative step, since they can be improved by implementing urban policies covering separative collection of waste.

<u>Combustion with treatment of smoke emissions.</u> VALORGA has updated a combustion line which is adapted for combustible waste (wood, plastics and fibres) and which presents the following two special properties : a high level of dry matter (70 to 75 %) and a regular composition. Combustion takes place in two very distinct phases :

- Pyrolitic type combustion in a revolving furnace; this enables accurate checking of temperatures up to 750°C, and a pre-neutralisation operation for the smoke emissions through slow mixing with the basic ash; other advantages are the absence of clinkers and economic maintenance.

Average composition of the ash

CO3Ca	15 to 18 %	
K2O	4 to 7 %	pH = 10.6
CaO	2 to 5 %	
MgO	1.5 %	

- A post-combustion operation for the gases from the pyrolitic sequence between 850 and 950°C for at least two seconds, whilst avoiding local overheating up to 1 200 to 1 300°C which results in the thermal decomposition of the water and the hydrochloric acid.

Neutralisation of the smoke. The operation takes place in a novel neutraliser which has been updated by VALORGA. The highly basic ash from the furnace is used to neutralise the sulphuric, nitric and hydrochloric acids. Spraying with this solution brings the temperature down below 170°C, and then a liquid gas film contact (in 3 stages) enables the neutralisation of smoke and the fixing of the escape of ash. At 110 - 120°C, the smoke is neutralised and contains steam; it is passed through a ring condenser, having a high output, greater than 90% on PCS, with remarkable latent heat recovery properties. This heat can cover the greater part of the energy requirements for the site, and any excess can be made available to industrial users (greenhouses, laundries, etc.). This unit makes it possible to perfectly manage the quality of smoke emissions and to comply with the new French regulations (9th June 1986) and with the future European regulations governing the incineration of municipal waste (Proposition COM (88) 71 final - March 1988).

Conclusion. The process developed by VALORGA is based on an industrial logic principle, at the same time as offering municipalities a technology which is flexible, adaptable to the deposit, recent but nevertheless reliable, thermically very efficient, and conveniently satisfactory as regards the latest standards concerning the environment (quality of smoke emissions, quality of ash, and agronomic quality concerning heavy metals). It offers an optimum valorisation for each fraction of domestic waste, depending on the local context and potential clients, and it could lead to the creation of poles of activity in the fields of agronomics and energy.

1. Diagram of the process

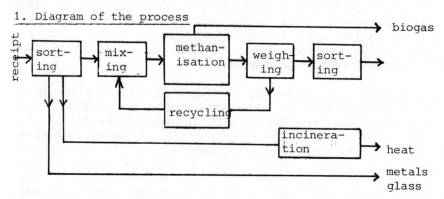

2. Gas network system

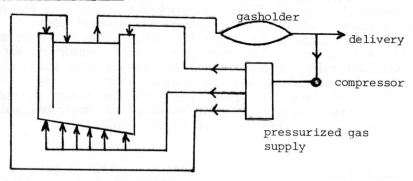

3. Diagram of discharge of matter (horizontal cross section)

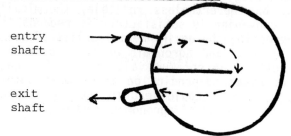

4. Profile diagram of production of biogas

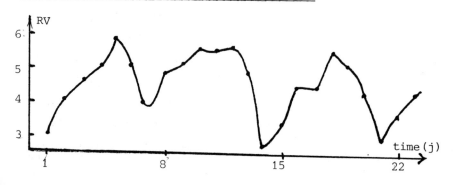

MICROBIOLOGICAL TREATMENT OF LIQUID AND SOLID WASTES OF URBAN SOURCE.
Dorronsoro,J.L.,Muñoz,R.,Sanchez,A.,Solano,M.L.,Torralba,B.
PROCESOS ENZIMATICOS S.A. Madrid, España.

INTRODUCTION

Urban waste represents a serious pollution problem,it can be an important source of organic matter and minerals if adequately managed.

Following the concept of compost defined by Parr (3), "the aerobic,thermophilic decomposition of organic wastes by means of mixed population of indigenous microorganisms under controlled conditions",the resulting product(compost) can be used as a soil conditioner and fertilizer when it is applied in appropiate doses.

When composting process is achieved,biologically transformed organic matter becomes a product free of bad odours, pathogenic microorganisms and phytotoxic substances.

As a consequence of urban wastes landfilling ,the action of rain , ground water,and chemical and biological decomposition, leachates are produced.If leachates are not properly controlled, they represent a very serious hazard to the environment because of its chemical and biological oxygen demand.

Spontaneous decomposition of organic matter in wastes occurs in nature,but these transformations are slow,discontinous and heterogeneous.

PROCESOS ENZIMATICOS,S.A. markets an enzymatic product from microbiological sources which acts as a catalyzer in the degradation of organic matter enhancing the composting process.

The aim of our research can be summarized in three points:
1- **Enhancement of the process and low energy consumption**
2- To guarantee a standard end-product safe for agriculture
3- To obtain a hygienic end-product safe for animals and humans

METHODS

Solid urban wastes

We have done three different experiments. In the first one, the urban waste came from the recycling plant of A D A R O (Madrid), and it was partially fermented. Piles of two metric tons were inoculated with our enzymatic product.

In the next two experiments, carried out in Cadiz, the residue used was fresh urban waste without previous fermentation. In this case, the composting was done on 25 ton piles. In one of them, sugar beet vinasse was added to fresh urban waste at a rate of 125 liters of vinasse to one metric ton of urban waste, and in the other one, just urban waste was composted.

Each test was done with two piles. One of them was treated with the enzymatic product at a rate of 1Kg of product to one metric ton of urban waste. The other one was used as control.

Composting was done in open air by the mechanically aerated windrow method (2). Piles were turned twice each week for 45 days, except in the third experiment, whose duration was 77 days because of climatic conditions.

Quality control monitoring included daily measurements of pile temperatures, periodic microbiological analysis and chemical analysis of the initial and the final product.

Leachates

The treatment of leachates from Pinto Landfill (Madrid), was carried out at laboratory scale in 1 l. flaks, in a rotatory shaker at room temperature.

Leachates were inoculated with different doses of product. The first one, 50%, was used in order to check the ability of product in degrading these leachates. Once the product proved to have a positive action, we decided to reduce the dose to a moderate 3%.

RESULTS AND DISCUSSION

	ADARO			CADIZ			
	T_0	T_f		T_0	T_f		
		Contr.	Inoc.		Contr.	Inoc.	Vinas
Total C	29.1	25	16.5	28.3	25.5	22.6	16.3
Total N	1.6	1.6	1.7	0.7	0.8	0.9	0.9
C/N	17.7	15.6	9.6	39.3	31	23.7	18.1
TOM	50.3	43.2	28.4	48.7	44	38.9	28.1
TOM/N	30.6	27	16.6	67.6	53.6	41	31.2
Index of Germination	0	60	98	0	41	100	99
F.Strep.10^2	5	0.9	0.4	5	0.4	0.2	–
Enterobac.10^3	300	20	0.2	200	50	0.2	2
F.Colif.10^2	10	1	1	60	60	3	1
Salmonel.10^2	10	–	–	20	20	–	–
Vibrios.10^3	200	70	0.2	10	1	0.7	7

Table I. Chemical and microbiological characteristics of starting material and compost.

The results obtained in the three different experiments carried out with solid urban wastes are shown in Table I.

C/N ratio

In the first experiment (ADARO), the C/N ratio observed in the compost inoculated with our enzymatic product was lower than in the controlling pile.

In the second and third experiments, carried out in Cadiz, a reduction of the C/N ratio was observed, but in this case, the percentage of reduction in the inoculated pile with regard to the controlling one was higher than in the first experiment.

Concerning the C/N ratio reduction in the experiments carried out in Cadiz, there is no difference between the inoculated compost and the inoculated pile mixed with vinasse.

TOM/N ratio

This ratio is used as an index of the maturity degree of a compost (1).

In the first experiment, the compost obtained with our product reached a higher maturity degree than that of the controlling pile during the same composting time; however, in the second and third experiments, the treated compost reached the maturity degree, while the controlling pile showed no signs of starting action.

The addition of vinasse improved the composting process.

Index of germination

In the three experiments, the index of germination of the **treated compost has been approximately 100%(5), while that of** the controlling ones oscillated between 40-60%.

Microbiological analysis

Table I shows a moderate decrease on the number of pathogen microorganisms in the controlling piles, but that decrease is **more evident in the compost inoculated with our enzymatic product.**

In this case, the end product proved to be free of pathogenic microorganisms: Salmonella free, fecal Streptococci less than 5×10^3 per gram and fecal Coliform less than 5×10^2 per gram (4).

LEACHATES TREATMENT

Experiments are actually carried out at laboratory scale in aerobic conditions.

The initial COD was 40,000 mg/l being the final COD 3,500 mg/l in the first dosification and 3,800 in the last one, which is equivalent to a BOD of 270 mg/l.

Leachates treated with enzymatic product present a decrease of pathogenic microorganisms higher than the untreated ones in both dosifications (Table II).

	$F C U ml^{-1}$		
	T_0	T_f	
		Control	50%	3%
F.Strep. 10^2	14	4.5	-	-
Enterobac. 10^3	300	800	-	2
F.Colirf. 10^2	20,000	10,000	4	5
Salmonell. 10^2	1,100	100	-	-
Vibrios .10^3	110	-	-	-

Table II. Evolution of the pathogenic microorganisms in treated leachates in comparison with a control.

BIBLIOGRAPHY

1- Godin,P. (1981). Fermentation et maturité des composts urbains,Compost Information.n6,2°tri.,2-3.
2- Gorby,H.B.(1960).Rapid composting U.S.Patient N2,947,619, assigned to Amer-Dross Disposal Corp.Date filed, March 22,1957;date issued,Aug 2,1960 (From Solid Waste Information System,Accession N°5902)
3- Parr,J.F.,Willson,G.B.,Chaney,R.L.,Sikora,L.J.,and Tester, C.F.(1978).Effect of certain chemical and physical factors on the composting process and product quality.In Proceedings of National Conference on Design of Municipal Sludge Compost Facilities,Hazardous Materials Control Researche.Institute Silver.Spring,Maryland.
4- Strauch,D.,and De Bertoldi,M.(1985).In Processing use of organic Sludge and Liquid Agricultural Wastes (Ed.P. L' Hermite)178-191.Academic Publishers,Dordrecht,Boston, Lancaster and Tokyo.
5- Zucconi,F.,Forte,M.,Monaco,A.,De Bertoldi,M.(1981).Biological Evaluation of Compost Maturity,Biocycle.Jul-Aug, 27-29.

Earthworm-filters for treatment of liquid sludge from fishfarms

I.W. Koster, D.A. Vroon, J. Bovendeur
Agricultural University, Department Water Pollution Control, De Dreijen 12, 6703 BC Wageningen, The Netherlands

Introduction

The high density culture of fish in systems characterized by application of water recirculation instead of water flow-through is a small, but rapidly growing bio-industry in The Netherlands. Water recirculation minimizes the energy requirements for maintaining the water temperature. At present a high density fish culturing system consists of three unit processes [1]: 1) growing of the fish in a fish tank; 2) purification of the water to be recirculated in a biofilter; 3) separation of suspended solids from the water in a lamella separator. The main discharge from such recirculation fish culturing systems is the effluent of the lamella separator. This sludge represents a potential environmental problem. In eel culturing the sludge discharge is 2.6 population equivalents per kg of daily feed input; in catfish culturing this is 3.2 [2]. Since the main fraction of the sludge are solids (uneaten feed, fish manure, excess biomass from biofilter) physical treatment can reduce the pollution discharge with more than 90 %. In this paper worm-filters, in which physical separation of the solids is combined with biological conversion of both solids and dissolved compounds will be discussed. Worm filters have already been applied successfully for treatment of liquid sewage sludge [3].

Methods & materials

A schematic representation of the filters is given in figure 1. At the start the top layer consisted of commercially available vermi-compost; during the experiments the sludge solids gathered there. Except for the blanks, the filters were populated with adult worms of the species <u>Eisenia fetida</u> who used the compost layer as habitat. Daily batch loads of sludge were applied during working days. Filters from which less than 50 % of the added liquid could be recovered within 24 hours were considered to be irreversibly clogged.

Sludge was obtained from a commercial eel farm. Its sludge discharge was thickened in order to obtain three grades (1, 2 and 3 % total solids).

Table 1: Average composition (± standard deviation, n=3) of the sludge per 1 % of total solids			
pH	approx. 6	nitrate-N	0
Kjeldahl-N	705±100 (mg/l)	nitrite-N	0
COD	13.9±1.2 (g/l)	ammonia-N	134±29 (mg/l)

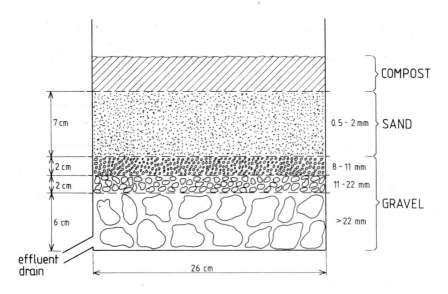

Figure 1: Schematic representation of an earthworm-filter

All analyses on effluent, sludge and compost were done according to standard methods of the Dutch Normalization Institute. The amino acids content of the worms (hydrolyzed in 6N HCl at 105 °C for 24 hours) was determined on a Biotronik LC6000E amino acids analyzer.

Results and discussion

function of the worms. The main function of the worms is to maintain the porosity of the filter. A worm-free filter operated at a loading rate of 0.5 kg COD/(m^2.week) with 1% TS sludge was clogged in the fourth week of operation, whereas a similarly operated filter with 100 worms at the start functioned well during the whole 12 week testing period. A filter with 100 worms at the start and loaded at 1.5 kg COD/(m^2.week) with 3%-TS sludge was clogged after 3 weeks of operation, whereas a similarly operated filter with 400 worms at the start remained open during its 6 week testing period. In all filters operated at loading rates exceeding 0.5 kg COD/(m^2.week) the worms could not consume all of the sludge input, which resulted in a visually noticable solids build-up.

effluent quality. The effluent quality varied with the loading rate. As an example of the effluent quality that can be achieved with worm-filters the average effluent quality of the filter with the highest loading rate tested is shown in table 2.

Table 2: Average effluent quality ± standard deviation (mg/l) of a worm-filter loaded at 1.5 kg COD/(m^2.week) with 3%-TS sludge			
pH	7.29±0.30	nitrate-N	89.7±20.1
Kjeldahl-N	263.6±74.3	nitrite-N	54.9±31.5
COD	1210±143	ammonia-N	217.6±60.9

nitrogen mass balance. A typical example of a nitrogen mass balance over a worm-filter is shown in figure 2. It appears that worm-filters are efficient in removing nitrogen. The average nitrogen removing efficiency of six filters started with 100 worms and operated at loading

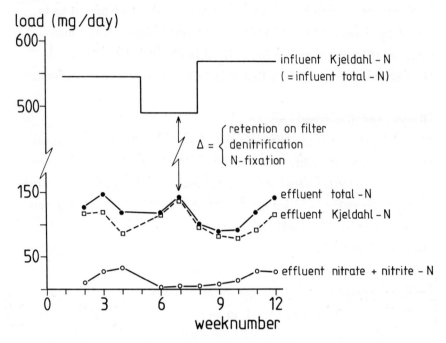

Figure 2: Nitrogen mass balance for an earthworm-filter operated at a loading rate of 1.0 kg COD/(m^2.week).

rates in the range 0.25 - 1.0 kg COD/(m^2.week) was found to be just over 70 %. For the filter of which results are shown in figure 2 it could be calculated that approximately 40 % of the nitrogen input remained in the top layer of the filter without being mineralized. Although the filter had a net production of worm biomass, this hardly accounts for N-fixation. Assuming that for any 100 mg of COD conversion 5 mg N were fixed in bacterial growth, the growth of bacterial biomass accounts for less than 10 % of the loss of mineralized nitrogen. This means that nitrification with subsequent denitrification caused more than 90 % of the removal of mineralized nitrogen.

COD mass balance. All worm-filters appeared to be highly efficient in removing COD from the sludge. Even the most highly loaded filter (1.5 kg COD/(m^2.week)) achieved a COD removal efficiency of 98 %.

However, the amount of COD that was actually oxidized by worms and/or bacteria varied widely with the loading rate. At 0.25 kg COD/(m^2.week) 98.5 % of the COD-input was oxidized, whereas at 1.5 kg COD/(m^2.week) this was only 49 %.

production and quality of worms. In all filters reproduction of the worm biomass occurred. The experimental period was too short to obtain valid information about the full production potential of the filters. The quality of the worms as feedstuff is probably good [4]. This is illustrated by the excellent amino acids composition of the worm-protein (table 3). The worms contained approximately 70 % protein on a dry weight basis.

table 3: Amino acids composition (g/100 g protein) of the protein of Eisenia fetida cultivated on fishfarm sludge

arginine	6.1	lysine	6.4	threonine	6.3
histidine	1.7	methionine	1.3	valine	5.8
leucine	6.8	isoleucine	5.2	phenylalanine	4.8

Tryptophan decomposed during sample preparation, so it could not be determined.

References

1. Bovendeur, J., Eding, E.H. and Henken, A.M. (1987) Design and performance of a water recirculation system for high-density culture of the african catfish, Clarias gariepinus (Burchell 1822), Aquaculture, 63, 329-353
2. Heinsbroek, L.T.N. (1987) Waste production and discharge of intensive culture of African catfish, European eel and rainbow trout in recirculation and flow-through systems, National report of The Netherlands for EIFAC working party on fish farm effluents, The Hague 29 May -1 June
3. Loehr, R.C., Martin, J.H. and Neuhauser, E.F. (1985) Liquid sludge stabilization using vermistabilization, J. Water Pollut. Control Fed. 57, 817-826

4. Stafford, E.A. and Tacon, A.G. (1984) Fish thrive on a diet of worms, Fish Farmer 7(3), 18-19

Acknowledgements

We would like to thank H. Abbink of Abbink's earthworm nursery at Woudenberg, The Netherlands for the gift of worms and vermicompost. We would also like to thank the family Van Zanten, who were of great help in collecting the sludge at their eel farm. Thanks are also due to W. Roelofsen of the Dept. of Microbiology for his help in the amino acids analyses.

Anaerobic Fermentation of the Vegetable, Fruit, and Yard Waste

Hans - Henning Spendlin, Rainer Stegmann

Technical University of Hamburg-Harburg
Hamburg, West Germany

1. Introduction

Separate collection of vegetable, fruit, and yard waste (vfy waste) will be practised more and more in West Germany, to reduce the landfilling waste amount. The share of vfy waste at West German municipal solid waste amounts to 44 %. This **separate** collected fraction will be composted at the moment. An **alternative** to this will be anaerobic fermentation of vfy waste including a short post composting phase. The advantages of anaerobic fermentation may be:
- less space requirement
- less odor problems
- shorter treatment time
- **positive** energy balance.

The end products are an energetic utilizable gas and a soil fertilizer, which is comparable with compost.

2. The anaerobic fermentation process

During anaerobic fermentation process organic material will be degraded biochemically in an airtight fermentor to a substrate, comparable to compost, and to biogas ($CH_4 \cong 60\%$ / $CO_2 \cong 40\%$).

The anaerobic fermentation is used since decades in sewage sludge treatment with good results. This process can't be applied easily to fermentation of vfy waste, because the process is designed for liquids or sludges with dry solid content < 10%. The vfy waste however has a dry solid content of about 60%.

That is why in some processes of the anaerobic fermentation vfy waste will be liquefied, that means to prepare the waste for pumping and stirring by shredding and adding water up to a watercontent of about 90%. Other **developments** operate on a so called dry fermentation process with a watercontent of about 60-70%.

Most most fermentation processes operate in mesophilic range. The retention time in the fermentor is about 2-25 days and gas production rate is about 90-200 m^3/t depending on the pretreatment step. In general one can say that retention time will be cut down and gas production rate will rise with increasing energy input during waste pretreatment.

Different anaerobic fermentation systems of vfy waste will be developed at the moment. Except of the France "Valorga" system, there is only few practical experience available.

3. Laboratory scale tests

Up to now different tests were carried out in laboratory scale lysimeters with a volume of 290 l at 35^0C temperature (Stegmann, 1981/1987). The lysimeters were run as batch tests. The vfy waste in the lysimeters were neither shredded before nor stirred during process. The moisture content had been adjusted at 65-70%. Process water was pumped discontinously in circulation to get a good distribution of the hydrolysed organic fraction.

Fig. 1 shows the cumulative methane production of the first 5 tests.

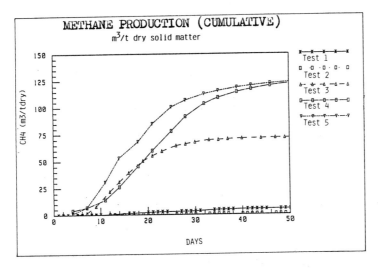

Fig. 1: Methane production (**cumulative**) test 1-5.
  Composition of the content:
  Test 1,2: vfy w.: compost = 1 : 1
  Test 3 : vfy w.: compost = 1 : 4
  Test 4 : vfy w.: digested
                    residue = 1,4: 1
  Test 5 : vfy w.: digested
                    residue = 1,4: 1 + $Ca(OH)_2$

While methane production from tests 1-3 starts to increase after a long period of time, results from tests 4 and 5 show that addition of digested residue (Lysimeter 4, 5) and addition of $Ca(OH)_2$ (Lysimeter 5) enhances the degradation process. Gas production rates give a good overview of biochemical degradation process rates in a fermenter (Fig. 2).

As a result of the tests, the following can be emphasized
- **separate** collected vfy waste can be degraded anaerobically
- main gas production rate and the maximum degradation rate lasts 20-25 days
- specific gas production rate is about 190 l gas/$kg_{dry\ solid}$ with a CH_4 content of about 60%.

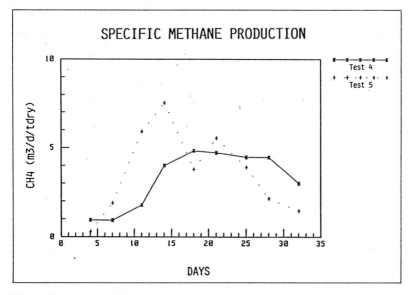

Fig. 2: Specific methane production rate - Test 4,5

At the moment we construct a plant in half technical scale ($10m^3$) to check laboratory scale results and to test complete operating technics (filling and emptying systems etc).

4. T.U.H.H. System

Based on good experience with a lot of lysimeter tests in laboratory scale to simulate anaerobic degradation processes, we started to develop a full scale system for anaerobic fermentation of vfy waste. We strove for the following targets:
- simple process engineering
- low water input
- low construction costs due to using of well known agricultural component parts
- low energy input.

Fig. 3 shows the process technic of the anaerobic fermentation plant developed at the Technical University Hamburg-Harburg. After a short optical inspection and manual picking out only obvious pollutions, the vfy waste will be sieved in a screening drum (Ø 5-10 cm). The sieve retention can be supplied again to the screening drum after another optical inspection and shredding or can be used as structure material to improve post composting.

The screenings will be mixed with degraded residue (ratio: 2/3 raw and 1/3 degraded material) and put into fermentor. Retention time in fermentor is about 20-25 days.

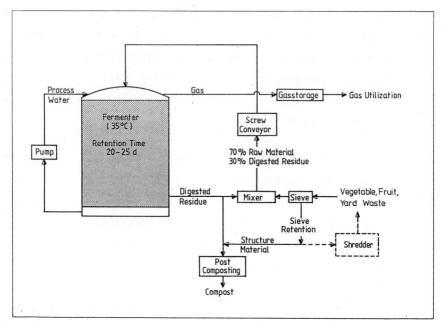

Fig. 3: Anaerobic treatment plant - T.U.H.H. system

The reactor will be a modified silo container from agriculture. It will be isolated and the temperature will be **held** at 35^0C. Filling and emptying systems will be executed gas tight. A definite amount of water will be added and recirculated.

Accumulated gas will be dried, stored and then used energetically. With the waste heat reactors will be warmed up.

The degraded residue will be mixed with structure material and post composted. The composting period is calculated with 1-2 **months** but this question can´t be answered so far.

Starting from the assumption that **separate** collection of vfy waste can seize about 5 kg/household/week (that's about 60% of vfy fraction in **municipal** waste), 1 m^3 reactor volume can degrade vfy waste of 25 households (corresponding 80 inhabitants).

5. References

Stegmann, R. (1981). Beschreibung eines Verfahrens zur Untersuchung anaerober Umsetzungsprozesse von festen Abfallstoffen im Labormaßstab, Müll und Abfall, No. 2.

Stegmann, R. and Spendlin, H.-H. (1987). Anaerobe Behandlung von Biomüll, VKS-Landesgruppe "Küstenländer", Lübeck.

Session 4

Sanitary Landfills

SCREENING TOXICANTS IN WASTE DEPOSIT RUNOFF

Timo Assmuth and Matti Melanen
National Board of Waters and Environment, Finland

Introduction

In order to control water pollution from waste deposits by toxic substances, they must be detected and assessed. For this an extensive chemical analysis may be needed. However, the appropriate contents, locations and timing of such analyses are difficult to define, given the multitude of potentially hazardous sites and toxicants emitted.

A broad preliminary screening of the various emissions is therefore often feasible. Screening may be defined here as characterization of emissions which integrates substances, effects, locations or periods, thus including e.g. chemical indicator and group analysis, bioassays and the collection of accumulated emissions.

As part of a 4-yr study of the environmental risks and clean-up needs caused by toxic wastes in Finnish landfills, 40 of them were investigated in 1986-87 with a screening approach. Some preliminary methodological results are described below.

Methods

Water samples were taken from runoff in landfill periphery and analyzed for key inorganics (standard methods), metals (ICP/AAS/GF), mineral oils (IRS), adsorbable organohalides (XAD and IC) and selected organics (GC). Mortality of <u>Daphnia magna</u> was measured with a simple 24 h test without dilutions, followed by the ISO standard test of LC_{50} values. Mutagenicity (Ames) to <u>Salmonella</u> was at times assayed. Additional background information assessment, hydrogeologic surveys and ground water and soil analyses were conducted.

Preliminary results and discussion

Chemical indicators of landfill impact

TABLE 1. The range of key chemical variables in surface water samples immediately downstream of the landfills and in the background (1986, AOX in -87).

Variable	Munic. fills min.-$\bar{x}$-max. (n=40-58)	Industr. fills min.-$\bar{x}$-max. (n=13-30)	Backgr. $\bar{x}$ (n=48-51)
Conduct., mS/m	5.6-157-620	5.6- 90-580	25
Ammonia, $mgNH_4$-N/l	0- 43-330	0-1.5-9.8	1.5
Sodium, mgNa/l	<1-110-600	1.3-168-700	30
Hardness, mmol/l	0.1-3.7-17.3	0.5-2.4-9.9	0.8
Iron, mgFe/l	0.4- 22-160	0.2- 21-480	2.8
Chlorides, mg/l	3.4-168-1500	4.0- 41-180	25
AOX, ug/l	<5-190-800	<5-350-3200	24

The runoff was dilute compared with leachates in other studies (Ehrig 1980, Kjeldsen 1986). Due to the variation between and within the sites, a combination of variables (e.g. conductivity, Na and Fe, at municipal fills also NH_4 and Cl^-) is often needed to indicate landfill impact (Table 1).

Except for Zn, heavy metals were rarely found with the flame methods. The highest concentrations were noted at industrial fills. Also mineral oil emissions were low, indicating retention or removal.

The elevated AOX and TOX values at many fills, esp. industrial ones, indicated organochloride emissions (cf. Kerndorff et al. 1985). TOCl and the sums of the analyzed specific organochlorides were correlated (Fig. 1). The latter at times exceeded the former; this may be due to losses (of volatiles) from the (sorptive) treatment. Even so, important toxicants may have remained uncovered.

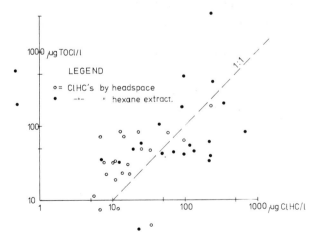

FIGURE 1. The relation between TOCl (AOCl and POCl) and the sum of specific organochlorides.

Toxicity of leachates as a screening tool

Acute toxicity was noted in 30 % of the downstream samples. Mean LC_{50}(48 h) in the standard test was only 45 %. The simple 24 h Daphnia mortality test was rather sensitive and correlated better with the chemical variables than the standard test (Table 2, Fig. 2). The two tests are mutually correlated, and the simple one may be used in fast pre-screening in acute toxicity assessment.

TABLE 2. The significant correlations (r values, n in exponents) of acute toxicity and some runoff quality measures.

Variable	Nr.	1	2	3	4	5
LC_{50} (48 h)	1	1.00				
Mortality	2	-0.71^{50}	1.00			
Conduct.	3	-0.48^{53}	0.74^{253}	1.00		
Hardness	4	-0.48^{53}	0.55^{253}	0.80^{253}	1.00	
COD	5	-0.72^{11}	0.78^{75}			1.00

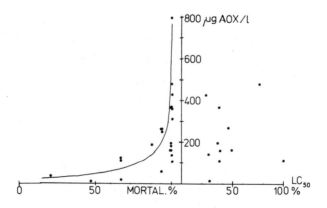

FIGURE 2. The relation between AOX and acute toxicity in 24 h tests and standard Daphnia tests.

Mutagenicity was masked by the acute bacterial toxicity of the samples but could be assayed with enzymatic activation, causing up to 3000 net revertants/l in the TA 100 strain.

The discrepancy between the biological responses and the chemical concentrations may be diminished if the array and precision of both kinds of analysis are increased and if the relative toxicity of each substance is accounted for. Unexplained residual variation in toxicity will still be present. This results from unknown toxicants (the cause for biotests in screening) but possibly from other factors as well, including artifacts ('wrong positives'). In addition, samples which appear non-toxic may contain otherwise harmful substances ('wrong negatives'). Bioassays or chemical analysis thus can not be used alone in screening hazardous emissions as effectively as in combination.

Integrated sampling approaches

Due to the variations of substance concentrations in space and time, integrated sampling must often be used in preliminary screening of toxicants. Time-integration is obtained with repeated sampling of runoff and with samples of substances accumulated from runoff either naturally (in sediments, soil and biota) or artificially (in incubated media or organisms). Spatial integration is needed e.g. in mapping the areal distribution of toxicants around a waste deposit. Surface and subsurface runoff routes must be covered; sampling of solid media may also be needed. True integration may be done by combining the individual samples for analysis. In preliminary screening indirect methods such as geophysical mapping techniques may be used which by their nature integrate areas.

References

Ehrig, H.-J. (1980). Beitrag zum quantitativen und qualitativen Wasserhaushalt von Mulldeponien. Veröffentl. Inst. Stadtbauwesen, TU Braunscweig. Heft 26. 392 p.

Kjeldsen, P. (1986). Attenuation of landfill leachate in soil and aquifer material. Ph.D. Thesis, Technical University of Denmark, Lyngby. 264 p.

Kerndorff, H. et al. (1985). WaBoLu-Hefte 5/1985. Bundesgesundheitsamt, Berlin. 175 p.

EXPERIENCE WITH NEW TECHNIQUES IN SLURRY CUTOFF WALL
CONSTRUCTION

Dr.-Ing. E. Biener, Dipl.-Ing. T. Sasse, HOCHTIEF AG, Essen
W.-Germany

1. INTRODUCTION

Vertical cutoff walls are an essential construction element
in landfill engineering and have gained increasing importance in both the construction of new waste disposal facilities and the rehabilitation of old landfills and contaminated sites. Due to their importance new techniques have
been developed in recent years to improve the quality of
cutoff walls. New cutoff wall compounds and methods of construction have increased the watertightness and chemical
resistance of cutoff walls against aggressive water. This
paper describes these new techniques in relation to the
Gerolsheim site and demonstrates the use of these techniques
taking into account the special site conditions of large
scale application.

2. DESCRIPTION OF THE SITE

Approximately 100 000 tons of hazardous waste are deposited
each year at the Gerolsheim hazardous waste disposal site.
The bottom of the disposal site has not been artificially
sealed and the majority of the waste lies upon a comprehensive watertight, quaternary silt horizon. In some areas, however, the waste pile lies directly upon sand and gravel
layers which effects the quality of the groundwater.

3. REMEDIATION CONCEPT

The remedial action concept for the waste pile required the encapsulation of the whole site. Investigations have revealed that a continuous tertiary silt layer exists beneath the sand and gravel horizon at depths up to 60 m. The main feature of the encapsulation is a cutoff wall which intercepts this watertight silt horizon. Together with the surface sealing, the cutoff wall and silt horizon form a closed system (Fig. 1).

4. SLURRY MIXTURES WITH HIGH PROPORTIONS OF SOLIDS

Compounds used for normal cutoff walls have, to date, been based on sodium bentonite, cement and water. This mixture has a specific weight of only 11.5 KN/m^3, with 90 % of the volume being water. The disadvantage of these low percentage solid mixtures (ca. 250 kg/m^3) is a limitation of the permeability coefficient to $k_f = 10^{-8}$ to 10^{-9} m/s and a limited resistance of the wall against aggressive seepage water.

As a part of a research and development project promoted by the Federal Ministry of Research and Technology (BMFT), the application of slurry mixtures with high proportions of solids (ca. 500 kg/m^3) was investigated for the Gerolsheim site. Laboratory developed these mixtures based on high proportions of calcium bentonite have permeabilities of 1 to 2 orders of magnitude less than those of the above mentioned compounds. These mixtures are also highly resistant to aggressive water attack. Following consideration of the laboratory research results, a slurry mixture, consisting of 8 kg/m^3 sodium bentonite, 306 kg/m^3 calcium-bentonite, 184 kg/m^3 blast-furnace slag cement and 809 kg/m^3 water was finally selected for large scale application.

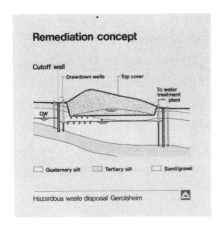

Fig. 1:
Remediation concept

Fig. 2:
Mixing plant

Fig. 3:
Cutoff wall execution

For processing reasons, specific materials such as super
liquifaction and retardation compounds of varying proportion also had to be added. Two mixing plants with completely
programmable control systems had to be set up for the production of these sophisticated slurry mixtures (Fig. 2).
The design of the plants had to allow for variable additions
of six different materials at any time.
Depending on the depth of the silt horizon, which the cutoff
wall had to intercept by 2 to 3 m, panels down to a depth
of 48 m were excavated by clamshells with improved grabs
and weights up to 15 tons (Fig. 3). In order to fully
guarantee the operational function of the cutoff wall, vital
importance was attached to the wall alignment. An electronic
measuring system developed especially for this application
was used for checking the position of each individual panel.
This system, which works in two dimensions, ensured that the
cutoff wall was built without any gaps and that the allowable
vertical deviations were not exceeded. This could be checked
at any time since measurement results **were** continuously
monitored.

5. DOUBLE LINING CUTOFF WALL

In some areas the calcium bentonite slurry cutoff wall had
to be supplemented by an additional synthetic sealing membrane. The advantage of combined sealant systems with respect
to watertightness, percolation of aggressive seepage water
and permeation of chlorinated hydrocarbons has been demonstrated through numerous tests and they are already being
used as bottom liners at new waste disposal sites.
To incorporate the synthetic membrane into the cutoff wall,
HOCHTIEF developed a special system which permitted the
placement of the liners, independent of weather conditions,

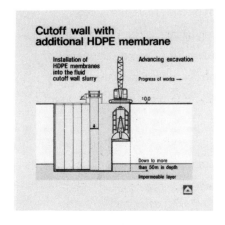

Fig. 4:
Sequence of work /
Combined sealant system

Fig. 5:
Placing of the HDPE-
membrane into the
cutoff wall

Fig. 6:
Back hoe excavating method
for cutoff walls

down to the necessary depths (Fig. 4). Specifically, the synthetic elements were lowered into the fluid slurry mixture, using a special drum which allowed continuous un-**winding** (Fig. 5). The individual elements **were** connected by a flexible locking system at the vertical edges, developed especially for this application. A double-chamber injection system renders the joint locking system tight. During work breaks, a special device shut off the locking system from the adjacent area and protected it against damage while excavation continued.

6. ALTERNATIVE EXCAVATION METHOD

At the Heßheim site, a domestic waste landfill just adjacent to the Gerolsheim site, HOCHTIEF used a technique developed in the United States for slurry cutoff wall construction. This technique, which has been rarely applied in W.-Germany, uses a back hoe for excavation instead of the normal cable operated clamshells. Although excavation depths are limited by this method, the joints in a wall can be minimized and a remarkable increase in production rate can be achieved. For quality control reasons, the location of the back hoe bucket was continuously controlled on a monitor by the operator during excavation. At the Heßheim site, walls with depths up to 12 m have been constructed using this technique (Fig. 6).

7. CONCLUSIONS

Using the above mentioned techniques, distinct improvements have been achieved in the quality of cutoff walls for encapsulating contaminated sites as well as for the construction of new waste disposal facilities.

DEPÅ 90 - PROGRESS REPORT FROM A SWEDISH R & D LANDFILL PROJECT

GUNNAR BERGVALL, National Environment Protection Board, Box 1302, S-171 25 Solna, Sweden

INTRODUCTION

When considering the need for landfill, it can be stated that in every waste management system, deposition plays a central role. There is no waste treatment system without deposition that can cope with all types of waste, and every other system gives rise in principle to residual products, which at present cannot be disposed of by any means other than deposition. Thus the landfill constitutes an essential basic resource in every waste management system. As regards the effects of landfill on the environment it might well be imagined that deposition, which has always been employed, would be well documented. Regrettably, it appears that information about the effects of landfill is extremely limited, especially as regards the important question of the effects of deposition in a long-term perspective.

It is true that significant R & D efforts have been made over the past years, but many problems remain. Further, there is no coherent picture of the landfill situation. In 1986 a program was therefore drawn up for carrying out a 5-year project with the aim of improving our knowledge about the effects of landfill on the environment, and the utilization of the resource potential of the landfill areas. The project was begun in 1987 and has been named Depå 90.

ORGANIZATION AND FINANCING

Depå 90 is a joint project sponsored by interested parties. It is led by a project board consisting of the National Environment Protection Board (SNV), the Swedish Association of Public Cleansing and Waste Management (RVF), the Swedish Environmental Research Institute (IVL) and the Joint Committee of Power and Heating Producers on Environmental Issues (KVM).

In order to achieve a broad basis within Depå 90 for the planning of **different** sub-projects and their implementation, as well as the results, there is a reference group attached to the project. The budget for Depå 90 amounts to about 1,3 million Swedish crowns per annum. The project is financed by the parties on the project board. In addition there are contributions from those responsible for the landfill areas, in connection with the application of the sub-projects at their particular sites. The magnitudes of these contributions vary from one sub-project to another.

AIMS OF THE PROJECT

The principal aims of the project are:
* To collect information which will form the basis for assessing the situation and lead to an acceptable environmental effect from landfill areas in a long-term perspective, under the present environmental conditions;
* To collect information which will form the basis for a rational utilization of the resource potential which is supplied to and present in landfill areas.

The aspect of Depå 90 which distinguishes it from other landfill projects, apart from the fact that this project aims to achieve a total picture of the deposition situation, is that the question of landfill is to be considered on a long-term basis. It should be obvious that landfill areas must be assessed in a very long-term perspective, over hundreds of years, since the landfill areas will exist/remain until the next Ice Age, or until other revolutionary events radically alter the environment. One important question for Depå 90 is to determine how long a period of the landfill area's existence will constitute a significant source of environmental disturbance.

The landfill areas are not only a source of environmental disturbance, but they also constitute a potential resource. Another important subject for Depå 90 is to investigate effective methods of utilizing this potential.

CURRENT PROJECTS

Opinions about the degree to which landfill affects the environment differ widely. One of the reasons for this is the lack of information about the effects of the landfill areas on the environment. Disturbance caused by landfill are largely concentrated to three pathways:
* gases which are emitted spontaneously
* leachate which runs off as surface water
* leachate which runs into the groundwater.

The intensity of the disturbances is a function of the breakdown conditions in the landfill.

An initial task for Depå 90 will be to describe the "normal" level for the different disturbance pathways. A combined assessment of these - a description of the environment - is an essential first stage and a prerequisite for evaluating and steering the direction to be taken by subsequent efforts within Depå 90.

A central task which Depå 90 must attempt to solve is the question of how long a landfill area continues to affect the environment, the magnitude of this effect, when protective measures can be discontinued, and how the landfill area should be formed during the period before the protective measures are discontinued and the landfill is left without any real supervision.

With the aim of throwing light on the above questions and goals, a number of sub-projects have been started. A brief description of some of the central sub-projects is given below.

Operating data

Basic information about the landfill areas will be obtained by collecting and processing the questionnaires which are sent out each year to about 40 randomly selected owners of landfill areas. This information deals with questions of waste quantities, landfill geometry, personnel requirements, leachate, gases, economics and other operational factors.

Breakdown

Experience has shown that the metal contents in the leachate are relatively low, in spite of the considerable metal potential in a landfill. This is attributed to various metal-fixating mechanisms such as sorption and sulphide precipitation, which occur principally in landfills that are in the so-called methane forming phase. The anaerobic parts of the landfill, principally the bottom layer, are assumed to function as an anaerobic barrier which regulates the leaching out of the metals.

Leachate

The physico-chemical composition of the leachate that runs off as surface water is considered by Depå 90 to be known in principle, due to the large number of investigations which have previously been carried out.

On the other hand, little is known about the toxic properties of the leachate. In one current sub-project the toxicity is being investigated, partly in respect of the variation of toxicity during the year and partly in respect of its variation from one landfill area to another. Eight landfill areas are included in this study. At the same time the physico-chemical characterization of the leachates is being carried out so as to facilitate the interpretation of the toxic studies.

In the case of the leachate which infiltrates and is transported in the groundwater, there is a lack of knowledge about how the leachate is stored in the groundwater aquifer and how different pollutants are transported in the groundwater. When investigating effects on the groundwater **it is** normally difficult to identify the polluted area, and thus to limit the number of observation pipes that have to be sunk. In order to be able to identify the polluted area, and thus sink the observation pipes with a greater accuracy, Depå 90 has tested geophysical methods of investigation. Thorough investigations of the groundwater situation at some landfill sites are planned, and these will apply **to some** of the geophysical methods.

As regards the treatment/management of leachate, an evaluation of the various ongoing leachate treatment projects is currently being made in order to find out which of the projects could prove interesting for possible cooperative efforts with Depå 90.

Combined deposition

The present Swedish view as regards the deposition of residual products from solid fuel combustion is that these should not be deposited together with "municipal" waste. However, there are advantages of a combined deposition of these two types of waste. In order to investigate possible differences in the environmental effects of combined deposition of municipal vaste and combustion residues, as opposed to the separate deposition of these two types of waste, a sub-project has been started. The investigations are estimated to take place during a three-year period at 2 different sites.

Gases

When organic wastes are deposited in a landfill, breakdown of the wastes takes place with the evolution of gases. These gases are energy-rich, and in recent years they have begun to be utilized for the production of energy. The time during which the gases can be utilizid for this purpose is very short by comparison with the time during which gas evolution/breakdown takes place in the landfill and the gases are released spontaneously into the atmosphere. The composition of the spontaneously emitted gases is hitherto unknown.

Depå 90 has carried out measurements of the spontaneously emitted gases using a new type of remote sensing instrument, DOAS, at four landfill sites. The measurements have been done for mercury, methyl mercury and benzene, and have been carried out continuously for between fourteen days and one month at each landfill.

Various sub-projects dealing with aspects such as literature surveys, gas quality, collected experience as regards the recovered gases, etc, are in progress or are planned.

Completion of landfill site

The way in which a landfill area is completed plays a major role in the long-term environmental effect of the landfill. Among the qualities that the completion must fulfil, it is quite obvious that the measures taken must last for a long time, they must provide as good a seal as possible so as to prevent percolation of precipitation through the landfill, in order to limit the pollution of surface and groundwater, and they must provide a good substrate for the **establishment** of vegetation. It may also be necessary for the completion measures to be arranged so that oxygen is prevented from penetrating into the landfill, so as to ensure the anaerobic barrier in the long-term perspective. One major difficulty when planning the completion measures is that subsidence takes place within the landfill masses, due for example to the breakdown processes in the landfill area. Plans are in progress for sub-projects which will throw light on this problem.

MEASUREMENT OF MERCURY VAPOR EMISSIONS FROM SWEDISH WASTE LANDFILLS

G. BERGVALL 1)
The National
Environmental Protection
Board, P.O. Box 1302
S-171 25 Solna
Sweden

R. KARLSSON AND
S. WALLIN 2)
Opsis AB
Ideon
S-223 70 Lund
Sweden

Mercury is a metal that occurs throughout the world in varying concentrations. Mercury can be disseminated in earth-, water- and airborne compounds. It can be transformed between various phases which can be spread in earth, water and air.

The atmosphere plays a major role in airborne dissemination of mercury between various locations. The flow of mercury from earth to air is substantially greater than the flow from earth to water. The flow from earth to air occurs principally through volatilization of mercury vapors, while the flow from earth to water is in the form of organic mercury compounds.

Production of the pure metal has increased in the 20th century. Until 1900, about 200,000 tons of mercury had been produced, while production in this century has reached about 500,000 tons to date. A large portion of this mercury is finally deposited in waste landfills.

To clarify whether mercury is emitted into the air from these landfills and in which quantities, measurements were made at four Swedish waste landfills between the spring and autumn of 1987.

DESCRIPTION OF MEASUREMENT SITES

1. Gärstad Landfill, Municipality of Linköping

The Swedish National Environmental Protection Board (SNV) authorized the Gärstad landfill on Nov. 20, 1973. The site had previously been used as a clay pit by Svenska Leca. Waste was first deposited at this site in 1974 and has consisted mainly of heavy building waste, non-combustible materials and excavated materials. Other waste such as slag and residual products of combustion as well as waste from slaughterhouses and latrines has also been landfilled here.

On occasions when the waste-combustion plant has been out of operation, the Gärstad landfill has also received waste that is normally burned. The water that leaches out of the landfill is handled in a water purification plant. All types of waste were landfilled at this site during a 5-year period when a new incinerator plant was under construction. By 1987, about 60% of dumping capacity had been utilized in the remaining available claypit. It will take about 2 years for available capacity to be fully utilized. At present, only non-combustible material is being landfilled, e.g. building waste and excavated material.

2. Blåberg Landfill, Municipality of Sundsvall

Norab Renhållnings AB has landfilled waste on the Blåberga site since 1970. All types of waste were deposited here until 1982. Hazardous waste was burned in special furnaces on this site while the Korstaverket power and heating plant was being built. Since 1982, waste has been ground and sorted at the Blåberg landfill and the combustible portion has been sent to Korstaverket. At present, only building waste and excavated material is landfilled at the site. About 70% of capacity at the current landfill has been utilized. The landfill covers 18 hectares, with a maximum fill depth of 20 meters. It is expected that this landfill can receive waste for another 2 years. The municipality plans to expand the landfill to obtain enough capacity to handle waste until 2000.

3. Isätra Landfill, Municipality of Sala

The Environmental Protection Board licensed this landfill in 1972. Waste landfilling started in 1973 and by 1985 an area of 7-8 hectares had been covered to an average depth of 6 meters. By 1987, another 8 hectares had been utilized at the same average depth of fill. About 90% of the landfill's capacity has now been utilized and an increase in fill depth to 10 meters has been proposed. The present site can be used for another 1.5 years. Expansion will allow operations to continue through 1997. Since 1984, household waste has been sent to Uppsala, where it is burned. The waste currently landfilled at Isätra consists of building waste and excavated material as well as non-combustible industrial and coarse waste.

4. Albäck landfill, Municipality of Trelleborg

The municipality was in charge of waste disposal at the Albäck landfill from 1953 until 1975-76, when SYSAV took over operations. All types of waste were landfilled here until 1983, but since that date waste has been sorted and the combustible portion has been transported to the SYSAV incinerator plant in Malmö. At present, household waste is landfilled at Albäck only if the incinerator plant is not in operation or is overloaded. The material landfilled at Albäck normally consists of building waste, industrial waste and excavated material. The volume of fill amounted to about 700.000 m3 in 1987, which corresponds to 70% of total capacity. It is expected that the landfill will be completely utilized by 1992. An expansion is being discussed.

MEASUREMENT TECHNOLOGY

Measurements were made with OPSIS technology, which is based on differential optic absorption spectroscopy. A specially designed transmitter sends a beam of light over the measurement path. A receptor focuses the light on an optic fiber cable which transmits it to the analysis unit, where it is broken up into spectra.

The light thus analysed has come in contact with various types of atóms and molecules on the path between the transmitter and the receiver. Specific types of atoms and molecules have the ability to absorb light of a specific wavelength or color. By determining which colors are not present in the analysed light, the presence of various substances in the measurement path can be shown. The concentrations of these substances can be determined by measuring the quantity of light that has been absorbed.

Concentrations are determined on the basis of the Lambert-Beer law, which states that:

$$C = \log (I'o/I)/(eL),$$

where C = concentration
I'o = light intensity without differential absorption
I = light intensity due to trace gas absorption
e = differential absorption cross section of the trace gas, and
L = length of absorption light path.

Measurement of atomic mercury takes place at 254 nm. Absorption of mercury can be affected by absorption from oxygen (Reference 1). This problem has been eliminated by evaluating O2 and Hg simultaneously in the spectral material.

MEASUREMENT STRATEGY

Concentrations of mercury vapors have been recorded as average values along different measurement paths. The paths have been positioned so as to enable measurement of the total flow of mercury from the waste landfills. The height of the paths was 1-2 meters above ground. Temperature, wind velocity and wind direction were recorded at a height of 5 meters above ground.

Mercury, temperature, wind velocity and wind direction have been recorded as average hourly values. The dates and places of measurements were as follows:

LANDFILL SITE	DATE
Linköping	May 25 - June 2
Sundsvall	June 26 - July 10
Sala	July 12 - August 6
Trelleborg	Sept. 21 - October 12

The measurement instruments were also calibrated for measurement of benzene, but no detectable quantities were recorded. The detection threshold for benzene is about 10 ug/m3. The mercury analysed occurred in the form of atomic mercury vapors. An attempt was made to analyse other forms of airborne mercury. However, the absorption cross-section for molecules is much lower than the absorption capacity of atomic mercury, so that the detection threshold was insufficient.

MEASUREMENT PATHS

SITE	LENGTH OF PATH, m
Linköping	290
Sundsvall	272
Sala	260
Trelleborg	252

ESTIMATION OF ERROR

Two types of measurement errors are possible:

1. Absolute errors resulting from faulty measurement of the light path and uncertain calibration of instruments.

2. Relative errors, caused by noise in the measurement signal.

Absolute errors

The instrument was calibrated with mercury vapors on the basis of steam-pressure data. The uncertainty of calibration is estimated at a maximum of +/- 5%. The uncertainty of the length of the measurement path is estimated at +/- 1%.

Relative errors

Noise in the measurement signal was less than 4 ng/m3.

MEASUREMENT RESULTS

Measurements of atomic mercury in the air above the landfill sites generated the following results:

Site	Hg content ng/m3	Average temp. C	Average wind velocity, m/s
Linköping	10.2	12.6	2.6
Sundsvall	19.2	15.8	2.2
Sala	23.6	15.5	1.7
Trelleborg	15.7	13.9	5.5

The measurements indicate the following back ground concentrations of mercury in each area:

Site	Hg content ng/m3
Linköping	4
Sundsvall	8
Sala	6
Trelleborg	8

CONCLUSIONS

Mercury vapors are present in increased quantities at the landfill sites where measurements were performed. There is a definite correlation between concentrations of mercury and ambient-air temperature. Higher temperatures lead to higher concentrations of mercury.

On the basis of the observations above, total emissions of airborne atomic mercury from an individual landfill are estimated at several kilos per year.

REFERENCES

1. Edner, H., Sunesson, A, Svanberg, S., Uneus, L. and Wallin, S. (1986), Applied Optics, 25 (3), pp. 403-409.

LEACHATE FROM LANDFILLS - A VALUABLE FERTILIZER FOR REVEGETATION OF LANDFILLS WITH GRASS OR ENERGY FORESTS

Torleif Bramryd
Department of Plant Ecology, University of Lund, Ecology Building,
Helgonavägen 5, 223 62 Lund, SWEDEN.

Abstract

Controlled landfilling of municipal solid waste can represent an environmentally attractive method to recover energy and other resources from the waste.
It is of greatest importance to prevent the transportation of leachates from the landfill to groundwater and surrounding surface waters. This will also help to create a reputation of landfills as "good neighbours".

This paper presents results from research investi- gations at a landfill in south Sweden, where the leachates were used as fertilizer in energy forests and grassplantations revegetating parts of the landfill area. The technique also represents an effective and economic way of treating the leachates.

Irrigation with leachate from the landfill significantly increased the survival and biomassproduction of a planted willow forest (Salix viminalis and S.aquatica) due to the supply of nitrogen, potassium, magnesium and other macro, as well as micro nutrients. Other research plots with grass (Dactylis glomerata) showed a near five-fold increase in production after irrigation with leachate.

Leachates from municipal sanitary landfills normally have low concentrations of most heavy metals and the uptake was very low both in willow and grass biomass; often lower than in the reference areas due to the higher biomass production.

The infiltration through the vegetation significantly reduced the concentrations of nutrients in the leachate.

1. INTRODUCTION

Leachates from landfills for municipal solid waste mostly contain large amounts of nutrients, especially nitrogen and potassium. Emissions of this leaching water to recipients without previous treatment pose a severe threat for eutrophication (increased growth of plant biomass with a decrease in available oxygen as a secondary effect). On the other hand the concentrations of heavy metals are in most cases small where the leachates could be regarded as a possible fertilizer for biomass production.

Extraction of methane from landfills as a "clean" energy source is an already established technique in many countries. However also the leachates from landfills could be turned into a valuable resource for producing still more energy from the landfilled garbage. These biological methods for energy production could probably during a 20 year period give over 80 % of the energy that would have been recovered by massburning of the solid waste. At the same time investment costs are very low and the environmental impact is significantly lower than for solid waste incineration.

Since 1983 experiments with leaching water irrigation are carried out in plantations with different willow and grass species at a landfill in Eslov, south Sweden.

The aims of the investigation are to evaluate the value of the leaching water as a fertilizer for energy-crops and to find an alternative way to purify the leaching water from nitrogen, potassium and other nutrients which might be potential threats to the recipients of the water.

2. METHODS

2.1 Investigation area

The fertilization experiments with irrigation of leaching water over willow and grass plantations were carried out at a landfill outside Eslöv in south Sweden. The landfill which started operation in the late 1940:s mainly consists of household garbage and other municipal waste. Only smaller amounts of industrial wastes and no chemical wastes have been brought to the landfill. The landfill is situated in a former peatarea where the amounts of incoming groundwater to the landfill is usually large. Table 1 shows the average chemical composition of the leaching water from the landfill. The heavy metal concentrations are very low and below the limits for drinking water (Hasselgren and Bramryd 1983).

In 1983 lysimeter studies were started with willow (Salix viminalis and S aquatica) and Dactylis glomerata grass. The exact amount of leaching water applied to the cells was determined as well as the amounts penetrating through the lysimeters. The chemical composition was determined before and after penetration of the lysimeters.

2.2 Sampling procedure and chemical analyses of soil and plant material

The leaching water was regularly analyzed and an average was calculated (table 1).

From each experiment unit 10-15 soil samples were collected from the upper 0-10 cm and the 10-30 cm of the soil and were put together to one sample for each unit. Sampling was performed at the end of the vegetation period. Stem wood and grass were harvested at the end of the vegetation period and analysed.

Table 1. Concentrations of elements in leaching water used for irrigation (mg/l).

Element	1983. X	1983. S.D	1984. X	1984. S.D	Element	1983. X	1983. S.D	1984. X	1984. S.D
Org-N	8.1	14	1.9	1.4	Cl	106	52	123	13
NH4-N	83	45	84	47	K	44	27	64	26
NO2-N	0.37	0.34	2.4	1.4	Fe	*		2.8	1.2
NO3-N	8.3	6.3	9.0	4.8	Cd	<0.001		<0.004	
Total-P	0.20	0.23	0.19	0.09	Cr	0.003	0.003	0.010	0.005
PO4-P	0.10	0.09	0.14	0.10	Cu	0.041	0.020	0.074	0.068
COD	190	155	190	46	Ni	0.032	0.026	0.025	0.011
BOD	17	11	17	9.0	Pb	0.035	0.031	<0.016	
pH	7.8	0.4	8.1	0.4	Zn	0.080	0.037	0.072	0.042

(*= no measurements done)

Ammonium and nitrate concentrations in the soil were determined according to an indophenolic method. After distillation of the ammonia and adding of a fenolate solution the intensity of the blue complex was determined colorimetrically. Nitrate was transformed to ammonium using a Devarda´s alloy powder. The pH in an aqueous solution was determined with a Metrohm digital pH-meter.

Kjeldahl-N in both soil- and biomassamples was determined with a distillation procedure after digestion with conc sulphuric acid.

Phosphorus was determined according to Kitson and Mellon (1944). After digestion in concentrated nitric acid the total concentration of P was determined colorimetrically after addition of a vanadate solution.

The concentrations of potassium, magnesium, sodium and calcium in soil samples were determined with a Varian Techtron AA6 atomic absorption spectrophotometer after extraction in an ammonium acetate solution.

Cadmium, chromium, copper, lead and zinc concentrations in soil sample were determined with an atomic absorption spectrophotometer after extraction with an EDTA solution.

The concentrations of K, Mg, Ca, Cd, Cr, Cu and Pb in biomass samples were determined with an atomic absorption spectrophotometer after digestion in concentrated nitric acid.

Biomass production of willow and grass was determined after harvesting and drying biomass from about 3-5 1x1 m plots.

3. RESULTS AND DISCUSSION

3.1 Biomass production

The annual biomassproduction of the willow plantation is presented in table 2. Irrigation with leaching water strongly increases the biomass productivity of both Salix viminalis and S aquatica. Fertilized plants have a productivity around 7-11 metric tons dw per hectare during the first year, compared to 0.3-1 tons dw per hectare in reference areas.

Table 2. Production of willow. One year old shoots on two year old rootsystems
Measurements after first and second growing season respectively.

Willow	Treatment	Age shoot/root (years)	Production (kg dw/m sq) stem-wood	leaf
S. viminalis	Leaching water	1/2	1.09	0.28
		2/3	2.63	0.59
	Control	1/2	0.11	0.02
		2/3	0.82	0.24
S. aquatica	Leaching water	1/2	0.65	0.19
		2/3	4.22	0.46
	Control	1/2	0.03	0.01
		2/3	0.59	0.07

The best biomass figures have been found for S aquatica. The annual biomass increment of two year old, irrigated root systems of S. aquatica were found to be up to 42 tons dw per hectare, while corresponding figures for S viminalis stopped at an annual increment of about 25-26 tons per hectare.

The results from this investigation are comparable with the best biomass figures obtained in willow plantations on heavily fertilized agricultural soils. The highest theoretically possible annual stem wood production has been calculated to approximately 30-50 metric tons dw per hectare and year (Linder and Lohammar 1982). The reference areas in this investigation only produced around 6-8 metric tons dw per hectare. The leaf biomass production during the second year is about 2-7 times larger in irrigated areas compared to reference areas.

If enough areas are available leaching water could support a significant biomassproduction. The produced biomass could either be burned directly or fermented to methane gas. Landfilling of municipal garbage in connection to methan extraction from the landfill and leachate utilization can be an interesting method to turn the garbage into a relatively clean energy source where most pollutants are bound up in the organic material in the landfill.

The annual production of the grass Dactylis glomerata has fluctuated between 6.5 and 21 metric tons dw per hectare in areas irrigated with leaching water while corresponding figures for untreated reference areas are about only 1.5 to 8.5 metric tons dw per hectare. Due to the relatively high water content in grass, anaerobic digestion for methane production seems to be the most relevant method to utilize the energy in grass biomass. Probably about 30-40 % of the biomass in dry grass could be converted to methane gas. According to Hasselgren and Bramryd (1986) the produced grass biomass in this investigation thus corresponds to about 15-20 MWh per hectare and year.

3.2 Effects on nutrient and heavy metal concentrations in soil

The concentrations of phosphorus in the top soil (0-10 cm deep) has increased significantly after three years of leachate-irrigation. Phosphorus is rather easily immobilized in the soil where the reduction of phosphorus from the leachates must be regarded as good.

The concentrations of potassium in the upper 10 cm of the soil is about 2-4 times larger after irrigation than in reference areas (tables 3a and 3b).

Table 3a. Metalconcentration and nitrogen in soilsamples from plots grown with Salix and grass. (μg/g dw if nothing else is mentioned)

Plot	Depth (cm)	Kjeld.-N (mg/g)	NH_4^+	NO_3^-	K	P	Ca	Mg	Na
					(μg/g dw)				
Willow									
Irrigated	0-10	4.90	2.86	1.79	830	308	4464	367	178
	10-30	3.42	2.58	33.3	313	171	3863	169	129
Not irrigated	0-10	3.03	*	*	209	110	3483	122	23
	10-30	1.62	3.51	*	83.0	60.0	2697	103	32
Grass									
Irrigated	0-10	3.27	2.52	9.89	371	526	3532	212	81
	10-30	1.65	2.08	8.47	127	327	2594	98	60
Not irrigated	0-10	3.64	4.24	0.72	701	254	3882	153	20
	10-30	2.30	2.71	28.8	89	175	3691	70	25

(*= below detection level)

Potassium is rather easily leached to deeper soil layers, where increases of potassium concentrations also was found in the layer 10-30 cm deep. Also the magnesium concentrations were somewhat larger after irrigation with leachates. Due to rather high clay concentrations phosphorus and most metals are relatively firmly bound up by clay minerals which decreases the transportation in the soil profile.

Nitrate concentrations have increased after leachate irrigation, especially below 10 cm, while the ammonium concentrations seem to be uneffected. During the prevailing pH a very rapid nitrification can be expected in the soil and only a minor part of the nitrogen is expected in an ammonium form. Nitrate is relatively easily leached, which explains the increased concentrations in the deeper soil layers.

The concentrations of heavy metals (table 3 b) are generally low in the leachates from a sanitary landfill with mainly household garbage (Hasselgren and Bramryd 1986). Thus the soil accumulation of heavy metals is very low after leachate irrigation and the natural variation exceeds the effects of the leaching water. In some cases the reference areas contain the higher concentrations of cadmium, chromium and lead.

Organic material in soil or in a landfill has a good capacity of binding up heavy metals. This has been shown in several previous investigations both in pilotstudies and in full scale (Bramryd 1985, Bramryd and Fransman 1985). Zinc and cadmium are metals which are relatively easily mobile in the soil due to a low capacity to form stable complexes with soil particles. Thus the accumulation of these elements is lower than for other heavy metals (eg Tyler 1978; Kerndorff and Schnitzer 1980; Bramryd 1980, 1983).

Table 3b. Heavy metal concentrations and pH in soilsamples from plots with Salix and grass.

Plot	Depth (cm)	pH (H2O)	Cd	Cr	Cu (µg/g dw)	Pb	Zn
Willow							
Irrigated	0-10	7.2	0.14	0.67	32.3	120	18.1
	10-30	7.2	0.14	0.30	22.8	57.3	8.00
Not irrigated	0-10	7.2	0.24	0.42	120	620	21.4
	10-30	7.2	0.05	0.50	20.0	117	10.0
Grass							
Irrigated	0-10	7.3	0.38	0.61	75.4	278	8.50
	10-30	7.7	0.29	0.26	46.8	178	38.7
Not irrigated	0-10	7.3	0.27	0.49	213	622	50.5
	10-30	7.7	0.24	0.52	180	802	54.6

The relatively high concentrations of lead in the top soil both in irrigated and reference areas can probably be explained by emissions from vehicles on the landfill. Lead usually forms very strong complexes with humus substances and clay colloides (Olson and Skagerboe 1975). Thus leaching of lead to deeper soil layers is a slow process (Siccama and Smith 1978).

Tyler and Westman (1979) has pointed out a strong connection between the soil microbial activity on one hand and the pH and metal concentrations on the other. The nitrogen mineralization is negatively effected at concentrations of copper exceeding 50-300 ppm (2-10 times the background) and at zinc concentrations above approximately 1200 ppm (6 times the background level). Thus it is of ultimate importance for the soil-vegetation system that the heavy metal concentrations aren't allowed to increase during the irrigation with leaching water. If this would be the case the irrigation must be decreased and the top soil layers should be replaced.

3.3 Effects on plant chemistry.

Due to a rapid growth and a fast transportation of nutrients and water to the upper parts of the plant the uptake of heavy metals soluted in the water will also be higher than in for example pine and spruce. Earlier investigations has shown that wood from Salix could contain up to 10 times as much of eg cadmium than wood material from eg spruce and pine. (Bramryd 1985).

Table 4. Heavy metal concentrations and nutrients in stems and leaf from willows and grass grown in plots irrigated with leach water. (μg/g dw if nothing else is mentioned)

Plot	Kjeld.- N (mg/g)	P	K	Mg	Ca	Cd	Cr	Cu (μg/gdw)	Pb
Willow									
Stem									
Irrigated									
S. viminalis	8.38	1.30	2.02	1.02	10.0	1.08	0.31	5.96	3.19
S. aquatic	7.42	1.45	3.16	1.14	14.4	1.24	0.31	5.15	1.12
Not irrigated									
S. viminalis	4.29	1.30	3.54	1.02	9.62	1.03	0.18	7.00	1.34
S. aquatic	3.56	1.41	1.77	1.26	13.6	1.00	0.31	5.56	0.95
Grass	D. glomerata								
Irrigated	24.5	2.63	5.49	2.91	10.6	0.12	0.66	9.80	3.51
Not irrigated	19.0	3.39	5.08	2,73	14.7	0.11	0.60	8.96	5.74

The uptake of phosphorus is larger in both Salix and Dactylis glomerata grass after irrigation with leaching water (table 4).

Recalculated as amount of phosphorus per square unit the differences will be even more evident due to the higher productivity of the irrigated plants. The uptake of heavy metals (table 4) is generelly low both in Salix and in Dactylis glomerata. Sometimes the uptake is larger in reference areas than in treated areas.

4. REFERENCES

Bramryd T 1980. Sewage sludge fertilization in pine forests - Ecological effects on soil and vegetation. In: Biogeochemistry of ancient and modern environments (Trudinger, Walter, Ralph Eds), Australian Academy of Science, Canberra, 405-412.

Bramryd T 1983. Uptake of heavy metals in pine forest vegetation fertilized with sewage sludge. In: Processing and Use of Sewage Sludge. (L`Hermite and Ott, eds). D Reidel Publishing Company. Dordrecht, 423-425.

Bramryd T 1985. Torv- och vedaska som gödselmedel. Effekter på produktion, näringsbalans och tungmetall- upptag. - Statens Naturvårdsverk PM 1997, 1-83.

Bramryd T och Fransman B 1985. Utvärdering av äldre gödslings- och kalkningsförsök med torv- och vedaska i Finland och Sverige.Statens Naturvårdsverk PM 1991.

Kerndorff H and Schnitzer M 1980. Sorption of metals on humic acid. Geochimica et Cosmoshimica Acta 44, 1701-1708.

Kitson and Mellon 1944. Ind Eng Chem Anal Ed 16, 379-383.

Linder and Lohammar 1982. Växter som solfångare. Forskning och Framsteg 7, 20-23.

Olson K W and Skagerboe R K 1975. Identification of soil lead compounds from automotive sources. Env Sci & Tech 9, 227-230.

Siccama T G and Smith W H 1978. Lead accumulation in a northern hardwood forest. Env Sci & Tech 12, 593-594.

Tyler G 1978. Leaching rates of heavy metal ions in forest soil. Water, Air and Soil Pollution 9, 137-148.

Tyler G and Westman L 1979. Effekter av tungmetallfororening på nedbrytningsprocesser i skogsmark; VI Metaller och Svavelsyra. Swedish Environmental Protection Board Report 1203.

Degasification of a capped landfill site for domestic refuse

by Heiko Doedens and Burkhard Weber, FR of Germany

1. Introduction

A waste disposal site is normally divided up in sections: one section with disposal activities and several completed sections. Completed sections can be covered with soils supplied from excavation on-site or delivered to the site without special requirements concerning quality. This type of a completed site is called **open** or **uncapped**. If a surface sealing system is applied to completed sections this type is called **capped or encapsulated**.

2. Uncapped waste disposal site for domestic refuse

Water plays an especially important role in landfill behaviour. Depending on the texture and the water content of the spread soil a certain amount of the precipitation infiltrates into the waste which causes the following advantages:

- Infiltrated water keeps a certain moisture content and so enhances the biodegradation of the organic compounds.

An uncapped waste disposal site is an uncontrolled anaerobic reactor and it has the following disadvantages:

- percolated water saturates the waste to field capacity, at that stage the amount of leachate can be 7,5 $m^3/(ha \cdot d)$ varying from 0 to 30 $m^3/(ha \cdot d)$; section in operation: on an average 5 $m^3/(ha \cdot d)$,
- the gas yield in german landfill sites compared to the gas production is less than 50 per cent,
- lost landfill gas migrates through the cover soil and hinders or injures the revegetation,
- gas-losses through cracks in the cover soil cause odour problems in the neighbourhood; minimum distances between landfill sites and residential quarters have to be more than 800 m if more than 250,000 t/a of waste are dumped.

3. Capped waste disposal site for domestic refuse

A capped waste disposal site is a quasi controlled anaerobic reactor and it has the following advantages:

- capping minimizes or prevents leachate production and thereby leachate treatment costs,
- gas-losses can be zero depending on the permeability of the chosen cover material, so the yield and quality of gas increases,
- gasmigration into the cover soil is prevented and the revegetation is protected,
- gas-losses originate only from the sector in operation; the minimum distance of 800 m is required when more than 400,000 t/a domestic refuse are dumped.

Disadvantages:

- encapsulation drys up the waste,
- stability of the slopes depends on the chosen sealing system and the gradient which should be 1:3 or flatter.

4. Capped waste disposal site with leachate recirculation

Cause the conservation of the waste in an encapsulated disposal site can not be the goal of waste treatment a leachate recirculation system has to be assigned for every capped sector. A recirculation ratio of 5 $m^3/(ha \cdot d)$ on an average - comparable to the leachate production of the landfill sector in operation - keeps a sufficient moisture content and so enhances the biodegradation.

5. Design and construction of the surface sealing system

The design and implementation of an effective capping-system involves first the selection of an appropriate cover material. The most common materials for capping are fine-grained soils such as clay or flexible synthetic membranes such as modified polyethylenes (PE or ECB). Clay as every soil has a certain porosity and so water and gases can permeate through it. Other disadvantages are possible cracking and shrinking and the thickness of the layer which should be at least 0.6 m which means a corresponding loss of disposal volume. PE-membranes are nearly impermeable, the required thickness of the layer is only 2 mm. Disadvantageous is the unknown long term behaviour of the synthetic material. Nevertheless PE or ECB is the recommended material for sealing especially when you compare the gas permeability of clay and PE (see fig. 5-1).

The design and construction of a top sealing system with a PE-membrane as a cover material is shown in fig. 5-2. For the sealing of slopes especially structured membranes are available. Grading should be 1:3 or flatter. Steeper slopes make special benches and trenches necessary to bind the membrane.

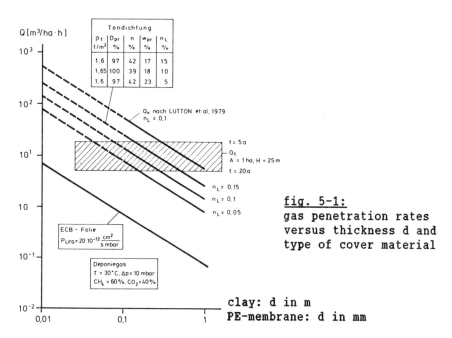

fig. 5-1:
gas penetration rates versus thickness d and type of cover material

clay: d in m
PE-membrane: d in mm

6. Pipe vent systems for degasification

Pipe vents consist of vertical or lateral perforated pipes usually surrounded by coarse gravel and discharging to a negative pressure collection system. Lateral systems can be installed and vented already during disposal activities. The gas

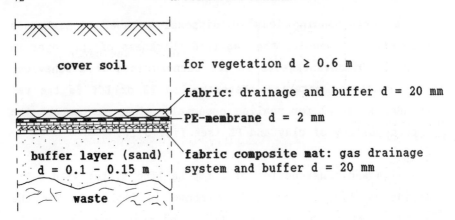

fig. 5-2: Surface sealing system

yield of vertical and lateral pipes in dependence of negative pressure, landfill-type and the suggested radius of influence is shown in fig. 6-1 and 6-2. To ensure adequate ventilation vertical pipes should be located 50 m apart in an uncapped site and 70 m in a capped site.

7. Degasification of the landfill site of Blankenhagen

The county of Northeim/FR of Germany has chosen a top sealing system as a remedial action for the landfill site of Blankenhagen. The 18 ha area of the waste disposal site is divided up in 5 parts, 3 of which are already capped at present. The construction of the surface sealing system is similar to that shown in fig. 5-2 with the exemption that the county of Northeim has chosen a liner system with 3 layers of 0.3 mm PE-membranes. Only for sealing the area around the gas-pipes and for the slopes they used 2 mm ECB-membranes. To keep the necessary moisture content leachate is being recirculated into the capped waste by injection wells and trenches. Vertical gas pipes and connecting lateral gas trenches are installed in the landfill already during operation. From special gas outlets (see fig. 7-1) which are bound up with the cover-foil the gas

radius R and catchment area A of vertical gas pipes

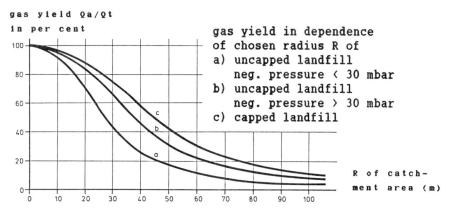

gas yield in dependence of chosen radius R of
a) uncapped landfill
   neg. pressure < 30 mbar
b) uncapped landfill
   neg. pressure > 30 mbar
c) capped landfill

fig. 6-1:   catchment area and gas yield for vertical gas pipes

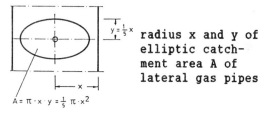

radius x and y of elliptic catchment area A of lateral gas pipes

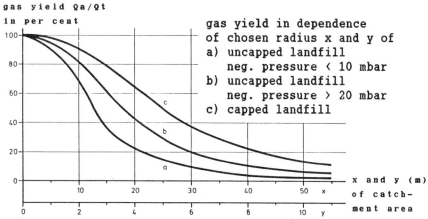

gas yield in dependence of chosen radius x and y of
a) uncapped landfill
   neg. pressure < 10 mbar
b) uncapped landfill
   neg. pressure > 20 mbar
c) capped landfill

fig. 6-2:   catchment area and gas yield for lateral gas pipes

is taken off in PE-pipes to a gas-suction plant. After passing gas filters, monitoring and security installations and a cooler the LFG will be delivered by a compressor via a 3 km PE-pipe line to a cement factory where the LFG will be burnt as a substitute fuel at a temperature of 1,400 °C. Excess gas will be burnt in a high temperature muffle furnace.

The expected gas production has been calculated on a personal-computer by a special mathematical model. The gas yield in proportion to the gas production turned out to be 60 to 70 per cent (see fig. 7-2). The analyzed gas concentrations were 52 to 56 per cent of volume methane and 34 to 39 per cent of carbondioxide.

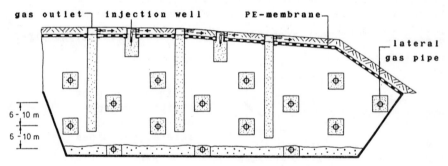

fig. 7-1: Encapsulated landfill site with lateral and vertical gas pipes and leachate recirculation system

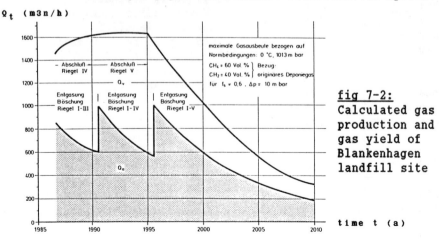

fig 7-2: Calculated gas production and gas yield of Blankenhagen landfill site

DESIGN AND EARLY RESULTS FROM THE LANDFILL GAS ENHANCEMENT
TEST CELLS AT BROGBOROUGH, BEDFORDSHIRE, UK

P Fletcher
Environmental Safety Centre, Harwell Laboratory, Didcot, UK

INTRODUCTION

Over the last decade there has been a substantial increase in the utilisation of landfill gas for its energy value. Worldwide there are now approximately 150 commercial schemes extracting landfill gas with a total energy value estimated as being in excess of 22 PJ per annum(1). Although extensive laboratory research has been undertaken to determine the factors which affect rates of waste degradation and hence gas production(2), very little work has been carried out under field conditions to see if rates of production or even total yields of gas may be improved. Hence, the Environmental Safety Centre of the Harwell Laboratory has been sponsored by the UK Department of Energy to undertake the project described in this paper.

PROJECT OBJECTIVES

The aim of this large-scale field project is to see whether landfill management techniques may be designed to enhance or optimise rates and yields of landfill gas production. Six different methods of placing/treating the domestic waste are being studied to observe the effect on gas production over several years. The particular experimental variables were chosen in the light of the results obtained from the earlier laboratory-scale studies carried out on incubated samples of waste. Where necessary, these variables (waste treatments) were adapted so that they can be generally applied to full-scale landfill operations. Thus it is hoped that the findings of this study may be used in the design and management of future sites.

EXPERIMENTAL DESIGN

Six large test cells have been constructed at an active landfill site in Bedfordshire, UK. One cell acts as a control and each of the others is being used to study the effect of a particular, distinct variable as follows:

- waste density;
- moisture content;
- temperature;
- co-disposal of sewage sludge;
- type of waste.

In order to be representative of normal landfilling operations, the cells were made sufficiently large to allow the site's usual waste delivery vehicles and compaction plant to be used in filling them. The cells also needed to contain sufficient waste to produce enough gas for the flow rates to be measured with adequate accuracy and, to be representative of a typical landfill, they had to be at least 10m deep. These considerations led to the nominal cell dimensions being set at 25m (width) x 40m (length) x 10m (depth) to give a total volume of 10,000 m^3 per cell. It was assumed that a placement density in the range 0.7-1.0 tm^{-3} would be achieved giving a minimum mass of waste of 7000 t in each cell. All the cells were filled in 2m lifts, again to be representative of standard landfilling practice. When the initial lifts of waste had been deposited it became apparent that a slightly lower placement density was being achieved and so an additional, sixth lift of waste was included to compensate for any shortfall in the total mass of waste upon completion of the cells.

The individual cells were isolated from each other and the rest of the waste in the site by means of substantial clay bunds. These were built up with the waste in 2m, interlocking, lifts. Waste was placed around all sides of the cells to provide support to the bunds and to ensure that there is no marked temperature gradient across the bunds.

A herringbone gas collection system was installed in the top layer of waste in each cell prior to capping the cells with a 1m thick clay seal. A plan of the test cells is presented in Figure 1.

CELL EXPERIMENTS

Cell 1, the control cell, was filled with domestic waste using the thin layer method of emplacement to conform with the recommended best practice(3). A minimal amount of granular material was used as cover between each 2m lift to

prevent windblown litter but allowing downward movement of
liquids and upward movement of gas.

A lower placement density was achieved in cell 2 by reducing
the degree of waste compaction. Waste was deposited in
single 2m lifts rather than the thin layers used to make up
each 'lift' in cell 1.

Cell 3 was filled in the same manner as the control cell but
a higher moisture content will be imposed by adding water or
by recirculating leachate.

The waste temperature will be increased in cell 4 by
injecting limited quantities of air into the lower levels.
This will promote a certain degree of aerobic decomposition
which is a highly exothermic process.

Sewage sludge was intimately mixed with the waste as lifts
2-6 were being deposited in cell 5. The sewage sludge should
provide additional nutrients and bacteria to increase the
rate of waste degradation.

Cell 6 was filled with commercial and non-hazardous
industrial waste in combination with domestic waste (50%) to
reflect more accurately the mixture of wastes accepted at
most UK landfill sites.

INITIAL RESULTS

Probes installed in each cell recorded high temperatures
(50-60°C) shortly after waste deposition, indicating aerobic
decomposition as the entrapped air was being used up.
Temperatures then, generally, settled down to 30-40°C but it
is too early to determine whether there are any significant
differences between cells.

Samples of gas have also been taken regularly from each lift
within each cell and analysed by gas chromatography. Results
from these analyses confirmed that oxygen was rapidly
consumed leading to anaerobic conditions within about two
months of waste deposition. The displacement of nitrogen,
principally by carbon dioxide and hydrogen, has occurred at
varying rates. In cells 1 and 3 the concentration of carbon
dioxide was as high as 90% within 2 months but it took much
longer for the nitrogen to be displaced from cell 2. This is
probably due to greater void space within the waste, and

hence more entrapped air, because of the lower placement density of waste in cell 2.

Methane concentrations have gradually increased from the onset of anaerobic conditions and, in most cells, had reached a few percent within four months of waste deposition. Levels of methane in cell 5, however, have increased more rapidly and were up to 20% within the same time (see Figure 2). This is presumably due to direct microbial action on the sewage sludge and it will be interesting to see if the trend continues in the long term.

Gas compositions and waste temperatures will continue to be monitored and flow meters have now been installed to record the rate and total volume of gas produced by each cell.

REFERENCES

1. Richards, K. "Landfill Gas – A Global Review", 7th International Biodeterioration Symposium, Cambridge, UK, September 1987.

2. Emberton, J R. "The Biological and Chemical Characterisation of Landfills". Energy from Landfill Gas Symposium, Solihull, UK, October 1986.

3. UK Department of the Environment. Waste Management Paper No 26 – Landfilling Wastes, HMSO.

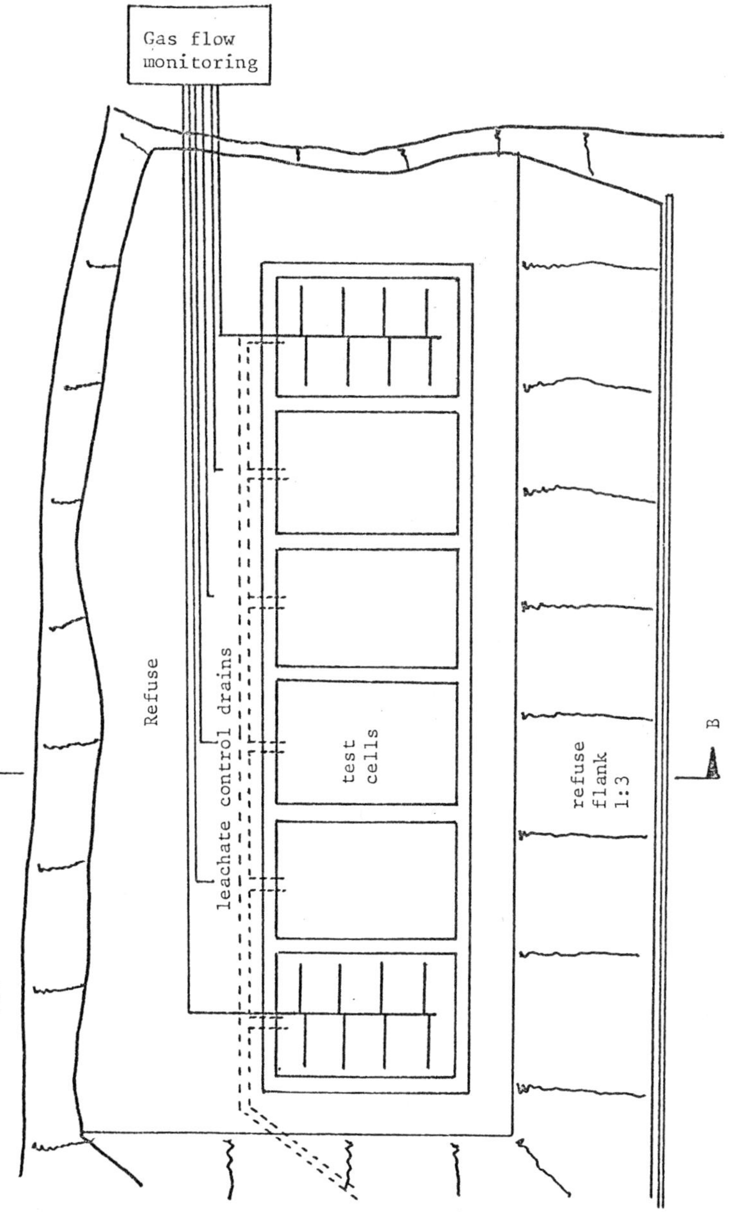

FIGURE 1 Plan of Brogborough Test Cells

Scale **1:1500** approx

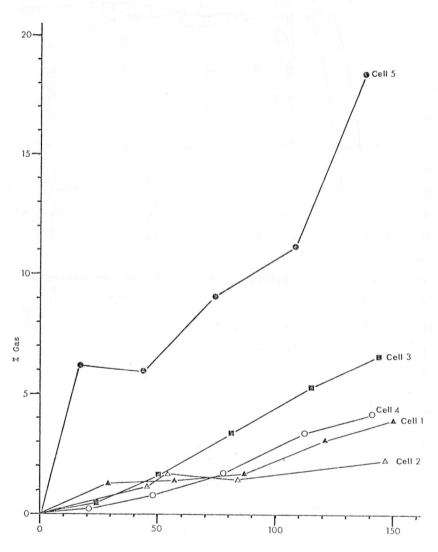

FIGURE 2

Brogborough Test Cells
Methane Concentrations, Lift 3

LEACHATE PURIFICATION BY IRRIGATING A PEAT BOG

Olov Holmstrand, Department of Geology,
Chalmers University of Technology

Bengt Troedsson, Scandiaconsult Väst AB

INTRODUCTION

Great efforts have been made to diminish the amount of wastes in Sweden. The waste treatment methods tested aim at making use, partly or fully, of the materials or the energy in the wastes, e.g. by incineration, composting or mechanical grading. However, the results have not been very positive. New environmental problems occur such as polluted flue gas from waste incineration. Still, a great part of the wastes will have to be dumped.

Although not harmless, landfilling seems to be a comparatively inexpensive as well as unavoidable waste treatment method. The most serious environmental problem is the production of polluted leachate. No completely acceptable method for cleaning leachate exists.

In Sweden landfills should be localized on impermeable ground to allow the leachate to be collected and treated (Statens Naturvårdsverk, 1974). Different methods for leachate purification have been tested. At **municipal** purification plants heavy metals end up in the sludge, which thus is made unusable for spreading on cultivated land, the consequence being the sludge has to be tipped on a landfill!

WASTE TREATMENT AT ÅMÅL AND SÄFFLE

The municipalities of Åmål and Säffle have together 30 000 inhabitants. The household and industrial wastes from the two and the household wastes from Grums (10 000 inhabitants) are dumped on Östby landfill, see Figure 1. Totally the present yearly dumping is 40 000 tons.

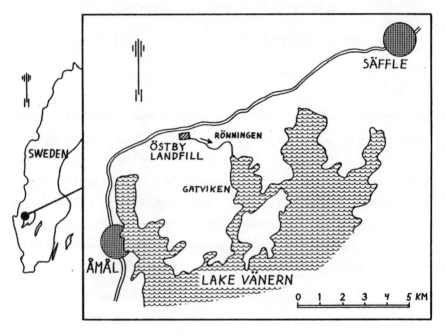

Figure 1. Östby landfill, location map.

The landfill is situated on a wooden, level mountain plateau. The bedrock consists mostly of granite, covered only with a thin layer of soil, mostly till. Originally the leachate **flowed** diffusely into the surroundings causing only local pollution of surface and ground water.

Legal permission for the landfill was considered in 1982. A **prerequisite** was that leachate should be collected and treated. The following alternatives were discussed:

1. Diffuse outflow into the surroundings. This "solution" had functioned rather well and could be described as spontantaneous irrigation. However, it was unacceptable.
2. Collection, storage and longterm **aeration** before outlet into the small stream Rönningen or to the lake Vänern. This was approved, but later considered not satisfactory.
3. Collection, storage and transmission to the sewage purification plant at Åmål. The long transmission pipe needed would be expensive and the result questionable.
4. Separate purification plant. Experiences from other places have shown that it is difficult and expensive to purify leachate in small, separate plants.
5. Local infiltration. The area surrounding the landfill is lacking suitable geological prerequisities.

Thus, none of these alternatives were technically and economically satisfactory. Instead, irrigation of a small bog close to the landfill was proposed, and also supported by the authorities concerned.

LEACHATE TREATMENT AT ÖSTBY LANDFILL

Leachate purification by irrigating a peat bog is based on the following assumptions:

- The leachate acts as fertilizer. The vegetation absorbs considerable amounts of nitrogen, phosphorus etc.
- Irrigation facilitates a more effective oxidation of oxygen-demanding substances.
- Peat has a high ion exchange capacity and a large specific surface, allowing high absorption of most leachate-substances.

- Irrigation of leachate on the vegetation will increase evaporation.

The irrigation plant was constructed in 1985-86 and put into operation in August 1986. A **principal** sketch is shown in Figure 2. Earlier investigations had proved that the infiltration into the ground can be neglected and consequently the entire runoff from the landfill area occurs as surface water.

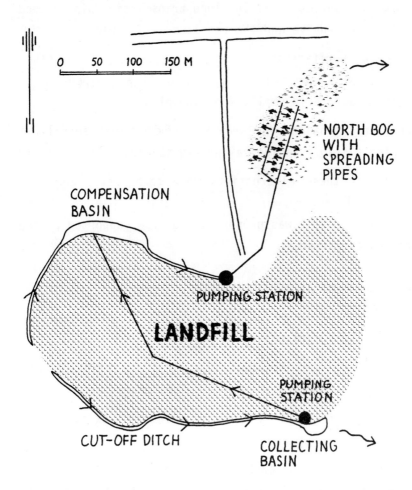

Figure 2. Östby landfill, simplified outline of leachate treatment.

The leachate from the landfill area is collected in surrounding cut-off trenches leading into two collecting basins. The bigger basin compensates flow variations and also is used for **aeration** by four submersible aearation pumps. An outlet ditch leads the leachate to the main pumping station. The ditch also forms a minor sediment trap. The leach water amount is monitored in the pumping station.

The area of the bog primarily used for irrigation is 6 000 m^2. The two spreading pipes are 15-20 m apart and approximately along the contour lines of the bog. The pipes are put on supporting frames 0.7-0.8 meters above the ground surface and thus also above the shrub-vegetation on the bog. The pipe material is PEH, diameter 0.1 m, with opposite spreading holes approximately on every third meter. The water is not sprayed by nozzles, thus avoiding the formation of inconvenient aerosols.

DOCUMENTATION AND EVALUATION OF THE METHOD

Thorough observations and documentation of the irrigation are demanded if a reliable appraisal is to be achieved. Unfortunately, so far the grants from the National Swedish Environment Protection Board have not been sufficient. However, the following main parts have been carried out:

- The project started with a theoretical preliminary study (Jansson et al., 1985). The main objective was to put together general conditions for irrigating leachate on peat areas and to give an account of similar installations in other places. The literature studied shows that peat has a considerable capacity for absorbing pollutants, that the leachate in Östby has a relatively low grade of pollution, and that no installation corresponding to the one in Östby has been described.

- Geology and vegetation have been mapped in detail within and around the bog. Also included was a small reference bog on the opposite side of the landfill. Peat samples have been analysed concerning the content of some metals.

- A network of 13 observation points has been established for sampling leachate, surface water and ground water. The water analyses comprise 16 physical and chemical parameters.

As mentioned the plant was put into operation in August 1986. The autumn was quite dry and thus the load was moderate. Sampling documented a good purification of the water leaving the area. The whole summer season 1987 was extremely wet. The sparse sampling shows that the installation obviously was over-loaded intermittently.

The irrigation plant at Östby should be studied for at least three years of operation before any conclusions can be drawn. If the method shows a satisfactory result, it should be useful at a number of landfills. Many municipal landfills in Sweden are situated at peat areas, which of course will facilitate the use of this method.

REFERENCES

Statens Naturvårdsverk (National Swedish Environment Protection Board), 1974: Localizing landfills considering geology and hydrogeology (in Swedish). Statens Naturvårdsverk, Publication 1974:24.

Jansson, B., Holmstrand, O., Troedsson, B., 1985: Leach water treatment by irrigating peat land, preliminary study (in Swedish). Statens Naturvårdsverk, Report 3045.

Disposal of Gypsum from Flue Gas Desulfusization (FGD) in land Reclamation

by

Lars Mikkel Johannessen,
Rambøll & Hannemann, Denmark.

Introduction

The Danish Government has passed laws to reduce sulfur dioxide emissions from power plants. As a result a power company in the Copenhagen area has decided to install FGD-wet scrubber systems with forced oxidation on some of their power plants.

The byproduct from the FGD-wet scrubber system is mainly gypsum, which is expected to be utilized in industry after purification. Unused gypsum and defective production runs meanwhile, have to be disposed of.

For that reason a land reclamation disposal site is being considered adjacent to the Copenhagen Airport. In Øresund, the area of disposal will be kept dry during the period of filling, by a drainage system in the surrounding embankments.

After final disposal it is intended to lead any leaching through the embankments, to the sea.

The main geotechnical test results of the examined disposal material and the main results of tests and theoretical calculations made to evaluate the environmental impacts from the planned disposal site, are given in the following.

Geotechnical properties

Gypsum is a porous material with a pore volume of app. 36% after compacting. It has hardly any strength and dissolves fairly easily. Therefore, gypsum has to be stabilized before being placed in a disposal site as land reclamation material.

Since fly ash is a puzzolan, and mixed with water and lime, a cementing material, it is an obvious step to add fly ash to FGD-gypsum as a stabilizer.

To assess reasonable geotechnical properties for the mixture, a test series was run with ratios between FGD-gypsum and fly ash of 70:30, 50:50 and 30:70 with and without lime addition (CaO).

Compressive strength tests show a substantial strength development in the first period of hardening, whereafter the development becomes more moderate. This process is basically related to the hydrate reactions.

The compressive strength increases with a higher content of fly ash in the FGD-gypsum. By adding 3% CaO to the fly ash stabilized FGD-gypsum, the compressive strength increases with a factor of 5 to 10, caused by higher hydrate reactions.

Permeability tests show a decrease in permeability with an increase in compressive strength. By way of the test series, a mix ratio of 70% FGD-gypsum and 30% fly ash with addition of 3% CaO, was found as a reasonable and practical disposal mixture. This mixture has a relatively high compressive strength of 6,000 - 7,000 kN/m^2 after 4-8 months of hardening. This gives possibilities for using the land reclamed as leisure area, parking space or other lightweight uses.

After final filling there will be no leachate collection from the site. With a permeability of the mix as low as 2.5×10^{-9} m/sek only a small percolation of water through the material is expected, with a minimum of leachate seeping out of the disposal site, as a result.

Environmental aspects

To assess the environmental impact of the planned disposal site on Øresund (the sea between Denmark and Sweden), leaching test series were carried out on the same mix ratio as mentioned above. The experiments were basically done to examine trace element concentrations in the leachate. The leachate was analyzed for the trace elements As, Ba, Cd, Cr, Hg, Mo, Se and Va.

The results show a decrease in the trace elements concentrations in the leachate, with a decreasing amount of fly ash in the stabilized FGD-gypsum. The results are not supprising, however, since tests with fly ash alone gives considerably higher concentrations of trace elements in the leachate than tests with FGD-gypsum without fly ash.

Experiments also showed that leaching performed with saltwater resulted in higher concentrations of trace elements than leaching performed with freshwater. As the disposed material is normally saturated with freshwater, and because of density differences between saltwater and freshwater, the leaching will mainly be caused by freshwater.

The trace elements concentration in the leachate from the stabilized FGD-gypsum decreases with an increase in the amount of water seeping through the material. This signifies a decrease in the concentration of trace elements in the leachate in the years after final disposal. The concentration of arsenic, however, will continue at an undiminished level for a longer period of time.

The maximum amounts of trace elements in the leachate from the disposal site per year, has been estimated for a filling time of 50 years and 25 years, see Table 1.

FGD-gypsum Disposal/year	10,000 m³	20,000 m³
Filling time / Trace element	50 years	25 years
As	0,63 (70)	0,63 (45)
Cd	<0,01 (41)	<0,01 (21)
Cr	27 (50)	50 (25)
Hg	<0,03 (41)	<0,03 (21)
Mo	15 (50)	23 (25)
Se	0,39 (49)	0,59 (25)
V	13 (53)	13 (28)

Table 1. Estimated maximum production of trace elements from the disposal site in kg per year. The number in brackets indicates the time of maximum production in years after the onset of filling of the site.

The trace element amounts are based on an annual percolation of 300 mm of precipitation through the disposed material. After the time indicated in the brackets, the amounts of trace element produced will decrease.

As appears in the table, the filling time is of great importance, especially for the maximum annual production of chromium. With the high concentration of Cr and very small concentrations of Cd and Hg, there is reason to believe that Cr is largely responsible for the toxicity of the leachate. Compared with experience from abroad and knowledge of the current conditions in Øresund, it is estimated that there will be only a minor toxic effect on the surrounding environment in a limited nearshore area, within less than 250 metres from the disposal site. In this area there is a possibility of trace elements accumulating in vegetation and organisms.

The discharge of trace elements from the disposal site will, to a great extent, accumulate in the deep zones of Øresund. Because the disposal site is located close to

Copenhagen and because sewage discharges in the area give a major contribution of trace elements to Øresund, the contribution from the disposal site will be insignificant.

Conclusion

The experiments have shown that it is possible to make a geotechnically stable FGD-gypsum disposal material with a mix ratio of 70% FGD-gypsum and 30% fly ash with addition of 3% lime. This mixture gives a relatively high strength material with limited deformations and a low permeability. The low permeability gives a small percolation through the material resulting in a small quantity of leachate.

Due to the small quantity of leachate, relatively low trace element concentrations and dilution in the sea recipient, the disposal site has only a minor impact on the surrounding environment. The conclusion must be that FGD-gypsum stabilized with fly ash has acceptable properties for use in land reclamation, as long as it is not placed near sensitive receiving waters.

IMPROVING THE EFFECTIVENESS AND RELIABILITY OF LANDFILL GAS PUMPING TRIALS

Dr B J W Manley, H S Tillotson, Environmental Resources Limited, London, England

1. INTRODUCTION

In recent years there has been growing recognition that gas generation from landfill sites can be both a potential environmental hazard and a marketable energy resource. Most gas recovery projects have been focussed on the latter aspect, i.e. exploiting the fuel value of the gas mixture. Relatively few schemes have been installed simply to control gas emissions and these are predominantly passive venting type systems.

Whilst the concept of drawing gas from a landfill site is simple, the mechanisms involved are complex and to date, control schemes have invariably been 'designed' by 'rules-of-thumb'. These schemes can often be ineffective, as the recent incident at Blue Circle's Stone plant in the UK has shown.

In order to design and specify effective gas control measures, a simple, low-cost site investigation procedure is required. Its aim should be to establish the true extent of off-site gas movement. Only when this has been examined should the design phase begin. Such a procedure should provide repeatability and be simple enough for non-specialist personnel to carry out.

2. GAS MIGRATION ASSESSMENT

The explosive potential of landfill gas has become a major issue and Her Majesty's Inspectorate of Pollution (HMIP) in the United Kingdom has circulated a letter to all Waste Disposal Authorities within England and Wales with the aim of initiating a responsive action to controlling gas emissions from landfill. In order to promote this initiative successfully, the Department of the Environment (DOE), through HMIP has identified the need for a means of quantifying the rates of gas migration in and around landfill sites.

When this has been achieved in a reliable and repeatable way, the actual means of intercepting hazardous gas flows may then be addressed.

At this stage of the exercise, geological factors are of great importance. For example, landfill sites situated in strata allowing the migration of gas (e.g. gravels, sandstones and fissured rocks) pose the greatest threat to the environment.

2.1 Site Evaluation Studies - Current Status

The mechanism of refuse degradation is fairly well understood. Pacey (Proceedings of joint UK, US Departments of Energy Conference 'Energy from Landfill Gas' Coventry, UK, 1986) has outlined a number of processes which predominate in anaerobic digestion and Emberton (Ibid) has examined the biological characteristics which govern this process. However, the kinetics are complex and as a result, mathematical modelling is not possible in a rigorous manner, particularly with respect to hazard mitigation.

The most appropriate method of obtaining migration data is to measure the behaviour of the landfill with respect to its surroundings. A common approach is to install a gas pumping rig which is fitted with flow and pressure monitoring equipment. The site is assessed in terms of:

- gas flow versus suction characteristics;
- abstraction "radius of influence", (the area from which the gas is drawn by the gas well), normally determined by monitoring at large distances from the well; and
- gas composition as a function of elapsed time from the start of pumping.

Information of this form is notoriously difficult to analyse owing to very large unquantifiable errors. An alternative method reported by the Building Research Establishment (1987) makes use of a simple surface surveying technique for measuring gas emissions to atmosphere. This requires sensitive methane detectors operating in their most sensitive range (ie at very low concentrations). Again the technique is unreliable because of large relative errors, caused by, amongst other things 'transients' attributable to changing weather conditions.

2.2 Migration Assessment by 'Migration Vector' Analysis

To establish reliable gas control, it is essential to form a detailed picture of the extent and nature of gas migration or movement. This requires a knowledge of the geology of the site as well as information on surface layer porosity. Off-site movement of gas almost always takes place. However, significant leakage is usually localised, following some sub-terranean fault, for example. These leaks or 'plumes' of gas, which can be described as migration vectors, can move many hundreds of metres. Rates of gas transport are normally small and these are controlled by diffusion.

ERL are developing a technique which aims to identify migration vectors whilst minimising the errors that limit the effectiveness of the BRE method.

The basis of the technique is to measure gas concentrations at shallow depths within the surface layers on and around a landfill. A grid coordinate system is used and measurements of methane concentration are made in 'drift holes' produced by driving a small diameter probe into the ground by hand.

In this way, readings of gas concentration can be taken for relatively large areas over short periods of time. These readings, as 'raw data', immediately highlight areas of risk and, therefore, complex data analysis is unnecessary prior to hazard control measures being implemented.

However, simple refinement of data is possible and a number of graphical presentation methods is possible. Examples are shown in Figures 1 and 2.

2.3 Technique Development

In order to produce a workable site evaluation technique, several issues require study.

o **Gas concentration measurement.**
This should evolve from an appraisal of available gas monitoring instrumentation and should take into account variations of CH_4/CO_2 ratios which may be caused by differential rates of diffusion.

o **Definition of gas flow pathways.**
Near the surface, these will be relatively uniformly distributed in the soil and sub-soil layers; at greater depths localised movement in discontinuities such as faults or fissures is more likely. (see Section 2.4)

o **Sampling Rationale**
Having considered methods of measurement and established likely gas flow fields, a sampling framework should be identified. This should aim to define the survey area, periodicity of sampling locations and depth and size of 'drift-hole' in which gas concentrations are actually measured.

2.4 Geological Considerations

Information may be derived from previously drilled boreholes, published geological maps or first hand observation. From such information the presence, extent and outcrop of formations conducive to gas migration may be identified.

A significant pathway of migration, even in supposedly 'impermeable' clay strata is fissure flow. Consequently a measure of the frequency, orientation and openness of jointing or fissures is essential in evaluating the magnitude of the potential migration hazard associated with this pathway.

This may be accomplished by modifying a technique used for some years in rock mechanics to assess the engineering behaviour of rock en masse. The presence of discontinuities in a rock mass determines its strength and stability. An understanding of the frequency and orientation of such discontinuities is therefore vital in the design of, for example, mines, tunnels and road cuttings.

Similarly a knowledge of the position, orientation and continuity of discontinuities in a rock mass surrounding a landfill is necessary to assess their effect on pathways and flow directions of gas moving away from a site.

The technique known as discontinuity or joint survey involves the visual mapping of exposed rock surfaces recording the, for example, type of structure, position in space, orientation, intensity; continuity, and discontinuity infilling (if any).

Results of the survey may be plotted on 'Rose' diagrams to indicate the direction or dip of the discontinuities and on stereographic projection or histograms to indicate the relative frequency of discontinuities in any particular exposure.

The development of this technique to gas hazard assessment is at present in its early stages but is showing promise.

3. GAS PRODUCTION RATE ASSESSMENT

The key to improving methodology is to reduce the relative measurement errors and to eliminate (if possible) 'random' factors which are inevitably associated with landfill sites. In principle, this can be achieved by examining the region very near to the point of abstraction.

Two alternative measurement variables can be considered:

o **Differential pressures.** These can be measured by installing small bore sampling tubes or piezometers at various depths within the landfill.

o **Gas surface flux.** The extent or rate of gas emissions at the surface will depend primarily on the gas permeability of the cover material. Gas fluxes can be determined by measuring rates of change of methane (or Carbon Dioxide) concentrations. From this type of information, an estimate or forecast of gas production rate can be made.

The analytical basis can be either:

- establishment of a pseudo-steady state for a particular gas abstraction rate; or
- step-response behaviour of a site to the 'forcing function' which is applied when gas abstraction rate is suddenly changed from zero to some predetermined value.

The step-response concept is novel and preliminary measurements of step-responses in terms of changes in methane surface fluxes are shown in Figure 3. No attempt has as yet been made to tackle the mathematics of these processes. However, a major simplification can be made by adopting a solution-by-analogy protocol. Two potentially suitable analogies have so far been identified:

- electronic and pneumatic process control design techniques, and
- unsteady state heat conduction.

The former could offer analytical solutions whereas the latter requires a finite difference-type numerical solution.

4. CONCLUSIONS

Controlling gas emissions completely can be a complex and expensive burden on landfilling operations. Nevertheless, control of gas is a prerequisite to the continuing acceptability of waste management by landfill.

In order to ensure absolute gas control, sites must be evaluated reliably, and quickly. To maximise cost-effectiveness the routine field work should preferably be carried out by non-specialist personnel.

This requires fundamental changes to existing methods of site investigation which are aimed primarily at the energy exploitation option.

Research and development work is urgently required to develop a study protocol which would receive universal acceptance throughout the public and private sectors of waste disposal operations.

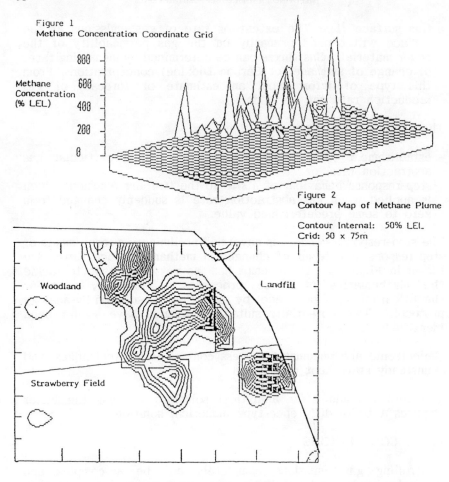

Figure 1
Methane Concentration Coordinate Grid

Figure 2
Contour Map of Methane Plume

Contour Interval: 50% LEL
Grid: 50 x 75m

Figure 3
Response to startup of gas pumping trial
shown by change in subsurface methane concentration

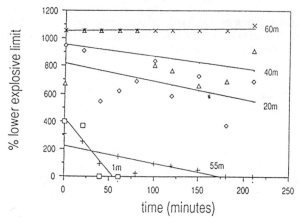

Three wells: gas flow
125 m³/hr
duration of test, 4 hrs.

All distances shown are
from Well A.

Enhanced Leaching and Decomposition of a Sanitary Landfill to Minimize Its Future Pollution Potential
A. Mennerich/K. Kruse [a], O. Schüttemeyer [b], R. Damiecki [c]

[a] Inst. f. Siedlungswasserw., TU Braunschweig
[b] Kreisverwaltung Düren
[c] Trienekens Deponiebetriebe

1. Introduction

The Landfill Horm of the Düren County, West-Germany, is situated in a former open ore mine. About $2.7 * 10^6$ m^3 of refuse were disposed off up to now. The final landfill volume will be about $7*10^6$ m^3. When landfilling was started in 1973, no sealing of the landfill base was provided. To avoid groundwater contamination, the water table is lowered beneath the landfill by pumping off about 400,000 m^3/a of groundwater contaminated with leachate, which is discharged to a sewage treatment plant. The mean concentrations of the contaminated water measured during 1986 are shown in table 1.

BOD_5	$(mg*l^{-1})$	100
COD	$(mg*l^{-1})$	200
NH_4-N	$(mg*l^{-1})$	50

Table 1: Mean concentrations of contaminated Water lifted in 1986

Because the whole landfill body will be below the groundwater level after closing up the landfill, the following sanitation measures are undertaken:

1. The top of the present "old" landfill body will be sealed by a double lining system, providing an appropriate landfill technique in the future.

2. Above the old landfill body (below the liner) a water distribution system will be installed to recycle a high amount of contaminated groundwater through the landfill body. The purpose of this intensive water throughput is an accelerated leaching of potential pollutants (Ammonia, heavy metals, **persistent** organics) and an enhancement of the anaerobic biological processes.

By this, it is intended to reach "near-groundwater-quality" of the leachate within a reasonable time. Then the groundwater lowering can be terminated.

The project is sponsored by the Federal Ministry of Environmental Protection, Bonn.

2. The Water Recirculation System

In cooperation between the Federal Bureau of Environmental Protection, Berlin the consultant and the Technische Universität Braunschweig a scheme for the water recirculation system has been developed. There were four inclined trenches laid up, reaching from the upper edge of the slope down to the intermediate sealing. Through those trenches, the distribution pipes for recirculation had to be laid. Control of the total water flow will be done by an automatic regulation consisting of a flow meter and an electromotive slide valve. Before the water is conducted into the recirculation system a pretreatment will take place by aeration to precipitate Calcium and Iron. A scheme of the water recirculation system is shown in Fig. 1. The detailed design of the system is still under construction.

3. The Stabilization of the Old Landfill Body

3.1 Goals of the Water Recirculation Projekt

The goal of the water recirculation is a stabilisation of the old part of the landfill, mainly by:

1. Biological processes
2. Eluation processes.

It is expected that both processes will be enhanced by an increased water throughput. It is quite difficult to predict the period needed to achieve a stabilisation, because this method was never practised before. Therefore, lab scale landfill simulators (V = 120 l) are operated to investigate the effects of the water recirculation.

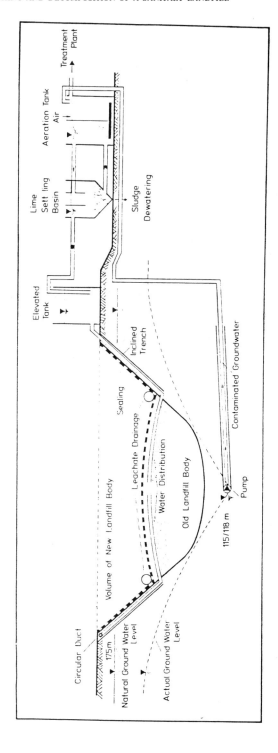

Fig. 1: Scheme of the Water Recirculation System

3.2 Lab Scale Investigations

To get preliminary informations on the space of time needed before reaching the stage of stabilization, lab scale landfill simulations are being operated at the Institut für Siedlungswasserwirtschaft. They are similar to those described by Ehrig (1986), Mennerich et al. (1986). The test setup is shown in Fig. 2. The test containers were filled with refuse which was won in drilling a gas well into the landfill body.

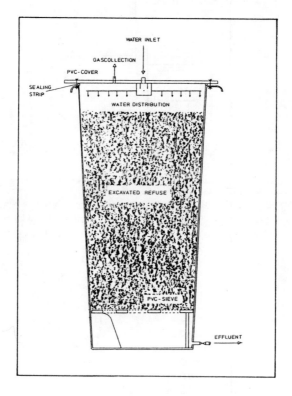

Fig. 2: The Test Container

Once to twice a week the water which has accumulated at the bottom of the containers (leachate) is withdrawn. Then, a certain amount of tap water is added to the containers via the distribution system. The gas produced is collected in gas tight bags. The quantities of gas and water are noted down and chemical analyses of gas and leachate are performed in certain intervals.

The operation conditions of the landfill simulators are shown in Tab. 2.

Container-No. (-)	Water added (l/Week)	(m^3/ha*d)	(m^3/t*y)	Operation Temperature (°C)
1	0	0	0	30
2	2	20	1.6	30
3	4	40	3.2	30
4	2x2	40	3.2	30
5	2x2	40	3.2	30
6	2x4	80	6.4	30
7	4	40	3.2	20

Table 2: Operation Conditions of the Lab Scale Test Containers

Because the tests are still running, a conclusive evaluation cannot yet be given. However, the following figures may give some interesting informations. The containers 3 through 5 behaved quite similarly and, **therefore**, are represented by one line (the solid one).

The patterns of the chloride concentrations observed in the leachate (Fig. 3) are merely influenced by the dilution rate, i.e. the water throughput. Hence, they can be used as a tracer. On the contrary, the COD concentrations (Fig. 4) are clearly affected by the anaerobic processes taking place in the refuse: In all containers, there were high concentrations between 4000 and 8000 mg*l^{-1} as COD to be observed at the beginning. This was due to initial acidification processes in the refuse, as confirmed by low methane concentrations in the produced gas at that time. The temperature was decisive for the time needed to reach the "stable methane phase", which is indicated through dropping COD-concentrations in the leachate: At 30°C, about 20 days were needed compared to 50 days at 20°C.

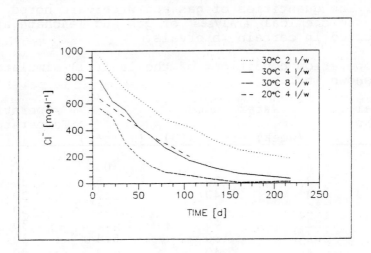

Fig. 3: Chloride

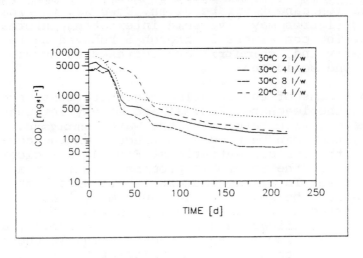

Fig. 4:: COD

On the other hand the curves indicate that after reaching the methane phase, the water throughput is the decisive factor for the COD-concentrations: The curves obtained at 20°C and 30°C with 4 l of water added weekly are very similar after 200 days of operation, while the containers with 2 l and 8 l water added per week yield higher and lower COD-concentrations, respectively. This would mean that, on a long term basis, the transport by percolating water will be the process decisive for the landfill stabilization even for organic matter: The phase of intensive anaerobic degradation processes is comparably short and leaves back refractory organic constituents, which can only be removed by leaching.

Another constituent also significant for the long term pollution potential of a landfill body is nitrogen. In Fig. 5, the patterns of the Ammonia-nitrogen concentrations present in the leachates are shown. Obviously, both the anaerobic biological and the dilution processes influence the amount of NH_4-N.: During anaerobic degradation NH_4-N is produced, which is again removed by the percolating water. As can be seen in Fig. 5, at an addition of 2 l water per week, NH_4-N concentrations in the leachate remain nearly constant. This means that there is almost an equilibrium between the released NH_4-N and NH_4-N removed. Even at 8 l per week the amount of NH_4-N released is large enough to prevent the "dilution curve" from approaching zero.

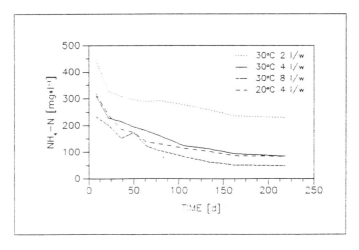

Fig. 5: NH_4-N

4. Conclusion

The lab scale tests indicate that the amount of water percolated through the Horm landfill will be decisive for the stabilization process, i.e. for the removal of pollution potential from the landfill body. The results of the lab scale landfill simulators cannot directly be used for the behavior of the real landfill: In lab scale, we get a "quick-motion" effect due to:

    Higher temperatures -> enhanced biological processes;

    Higher water addition per weight unit refuse -> enhanced eluation.

The scale up to landfill conditions requires some additional analyses and investigations, which are still under operation.

Ehrig H.-J.	(1986)	Untersuchungen zur Gasproduktion aus Hausmüll Müll und Abfall 5/86 S. 173-183
Mennerich A., Wolffson Ch.	(1986)	Überlegungen zur langfristigen Entwicklung und Kontrolle von Emissionen aus Hausmülldeponien Abfallwirtschaft in Forschung und Praxis, Vol.16 E.Schmidt Verlag Berlin
Stegmann R., Mennerich A.	(1983)	Entwicklung eines Testverfahrens zur gemeinsamen Ablagerung von kommunalen und industriellen Abfällen BMFT Research Report not puplished
Chian, De Walle, Hummerburg	(1985)	Effects of Moisture Regimes.... Zit. in Cord-Landwehr: Stabilisierung von Mülldeponien durch eine Sickerwasserkreislaufführung BMFT Research Report not published

LONG-TERM DISPOSAL FACILITIES FOR CHEMICAL WASTES
WITH RESPECT TO ENVIRONMENTAL PROTECTION

Parráková, E., Fratrič, I., Mayer, J.

1. INTRODUCTION

Present technologies of the chemical industry are far from meeting the desired characteristics of low-waste production. Thus, production of considerable amounts of wastes necessitates construction of long-term large-capacity refuse disposal sites outside the factory area. Prior to choosing the locality it is necessary to investigate the geological and hydrological conditions, to take into consideration the landscape protection, the necessity of land occupation, soil stabilization, and other parameters.

2. CASE STUDY

2.1 Local and geological conditions

The controlled disposal site had been constructed in the distance about 30 km from the factory on a flat ground, the area covering 34 ha. According to the hydrological investigation the territory **consists** of neogene and quarternary sediments. The neogene consists chiefly of varicoloured clays. Sands are rather rare, creating only small lenses. Quartenary sediments are formed by loams and silty clay loams, locally by sand and gravel. The chosen locality shows favourable conditions with respect to the thickness of quarternary sediments /mostly up to 15 m/, lower **occurrence** of precipitation, low concentration of population within the area, and especially due to the circumstance that the territory does not belong to the protected drinking water supply area. The subsoil permeability was determined by means of the filtration coefficient amounting to 10^{-8} to 10^{-9} $m.s^{-1}$, in clayey sands 10^{-6} $m.s^{-1}$.

2.2 Type of disposal site

In the case concerned the wastes are stored on the surface of geological formations, where the geological factor is decisive as a protective barrier between the body of the disposal site and the biosphere. Therefore, the geologic medium where the disposal site should be situated, should have high

sorption affinity and low permeability with respect to harmful substances. The subsoil under the disposal site should be sufficiently spacious.
In our case those conditions are met and thus the advantages of unloading in geological media, namely their protective factor, may be utilized. Though, certain disadvantages are also to be considered, as for instance: the poor **homogeneity** of the geologic subsoil, as well as the fact that it is not feasible to test the impermeability of the geological formations in the chosen locality, since even the tests would disturb their wholeness. From practical aspects one factor cannot be considered, namely the time.

Thirteen units 380 m long, 9 m wide, the space between dividing dams being 40 m, deep about 8-9 m, the slope being 1:2, are to be successively constructed on the demarcated area. The first unit served as experimental one after the wall colmatage with sodium silicate and cement had been performed. The second and remaining units would be sealed with welded plastic foils prior to waste unloading. The polluted rainfall water from the charged part of the unit is pumped and conveyed to the wastewater treatment plant. Clean rainfall water is pumped and used for irrigation and experimental fish breeding.

2.3 Quality of stored wastes

With respect to their chemical composition and consistence the unloaded wastes are rather heterogenous. They contain wastes originating from the production of chemicals applied in agriculture in plant protection, fertilizers, rubber plant wastes, arti**ficial** fibres, sludges from wastewater treatment plant, as well as sediments from various production branches.

2.4 Interaction of leaching water from the disposal site subsoil

It is evident that after a certain period the soluble substances start to penetrate into the adjacent soil and would probably jeopardize the groundwater quality within the close, and later on also within the vast environment. The penetration of leaching water is less expressive due to the spheric character of contaminants spreading, and resulting dilution. Advancement of dissolved substances through the soil is an extraordinary complicated process, during which the soil is acting as sorbent, ion exchanger, and coagulant. Thus, a fraction of waterborne substances is removed by the soil in dependence upon time and distance from the units. This occurs on the interfaces between the solid /soil/ and liquid phase /groundwater/, the processes being rather complex, since in the distance of several micrometers from the earth with alcalic reaction may be particles with acid reaction. Different are also the sorption properties of soil particles, in which prevails a considerable heterogeneity with respect to internal surface and field of force /1/.

3. EXPERIMENTAL PART

3.1 Monitoring of the leaching from the units into groundwater

Changes of physical-chemical characteristics of groundwater as influenced by leaching water are followed by means of analyses of groundwater samples, withdrawn from observation wells. The actual operation feasibilities of analytical laboratories show the necessity to determine the concentration of chosen indicators and not to trace all respective components. As one of the basic indicators may be considered the specific conductivity, reflecting the total concentration of dissolved matters, dissociable in water, and able to form cations or anions. This value is **remarkably** sensitive and quickly determinable; it has only one drawback, namely that it does not indicate the presence of in water soluble, though undissociated organic matter. Another group indicator value is the oxidability, including a sum of inorganic, though namely organic substances, capable to reduce powerful oxidizers. As far as anions are concerned chlorides, nitrates, sulphates, and phosphates, the first two forming in water well soluble salts, while the last two anions form very slightly soluble compounds within the soil. As far as cations are concerned ammonia, iron, potassium, sodium, calcium and magnesium /3/.

3.2 Evaluation of the effect of leachates from the units into groundwater

It is easy to perform this evaluation from the theoretical point of view: e.g. by means of a system of two coordinates, indicating concentration /y/ and time /x/. Concentrations measured in the course of experiment on the coordinate y are evaluated with

respect to the values determined prior to the construction of the disposal units. The time scale x is normalized by the distance to the well, by the average groundwater flow velocity within the porous media. Though, the situation is more sophisticated in practice. The background concentration is a conservative parameter, subject at least in the given case, to a rather high variation coefficient. Besides, in the course of the experiment the maximum obtained value of the studied parameter depends upon:
- the leachate concentration
- the concentration of the same component in uncontaminated water, and
- the spheric dilution of the component.

The turn in leaching into the observed wells may be attained by means of adsorption on soil, and the studied component would be detected with a time phase delay. The maximum leaching should not be demonstrated simultaneously in all studied parameters.

The advancement of the contamination peak from the disposal site to the control well can be traced using simple statistical tests and equations of one-dimensional transport of the dissolved matter, taking into consideration flow, dispersion and adsorption of dissolved matters. The equation can be solved analytically /2/.

The situation in the investigated disposal site is rather unfavourable with respect to the evaluation, since the units are situated within rather variable layers with respect to the lithology. The hydrogeological survey showed a complex of quarternary sediments having predominantly a thickness of 15 m. Under this layer occurs a continuous subsoil formed mainly by clays, having a permeability of 10^{-7} to 10^{-9} m. sec^{-1}.

In those formations layers of sand occur in certain areas, outcropping in places to the surface, where a refill of collectors with precipitation water can take place. Thus the variability of results in time is explained. Adjacent to the disposal site observation wells have been constructed reaching to the depth of 10-15 m below the terrain level. From the total of 42 drilled wells none of 22 wells showed water, which evidenced the irregularity of the geological formations. Physical-chemical parameters and their variations in time were monitored in 13 wells. Grouping of wells may be arranged according to the distance from the units or according to mineral composition and variability in time. The second classification affords a better image on the possible effects on groundwater caused by wastes stored in the unit. Firstly, the volume concentrations in mg.l^{-1} in the unit 1 /K 1/ are compared with the unit 2 /K 2/ /Table 1/.

Tab. 1. Average volume concentrations of parameters in mg.l^{-1} in K 1 and K 2

Parameter	Unit 1	Unit 2
Chlorides	245,3	10,9
Nitrates	5,0	2,1
Sulphates	778,8	126,2
Potassium	146,0	3,9
Sodium	438,0	22,2
Calcium	260,0	54,2

The water in K 1 is predominantly of sodium-sulphate character, and in K 2 of calcium-sulphate character. Arranging wells into groups according to mineral composition, waters showing no significant variations of the type of chemism belong to the first class, waters showing changes in chemism, especially due to sulphate ion, belong to the second class. The third class is formed by groundwaters, the chemism of which showed certain variability. The significance of contamination was established according to the t-value and was demonstrated in relatively few cases. In the first class only one from five monitored wells showed an increase in sulphates /t-3,2/, sodium /t-3,5/, potassium /t-5,8/, and in another well increased values of nitrates /t-4,1/, and calcium /t-14,5/. In the second class a group of four wells was studied; in three of them an increase was determined, namely in one well oxidizability /t-5,6/ and calcium /t-3,2/, in another one sodium /t-2,78/ and in the third one chlorides /t-2,99/. Four wells formed the third class, one well showing an increase in sulphates /t-2,92/, the other one in conductivity /t-6,6/, chlorides /t-3,1/ and sulphates /t-4,9/, calcium /t-4,8/, magnesium /t-4,5/ and the third one in nitrates /t-3,6/, sodium /t-8,5/, calcium /t-9,6/ and magnesium /t-7,9/. That means, that from 130 monitored values in 19 cases an increase of the indicator against the **background** was identified.

4. DISCUSSION

Studying the influence of wastes and of stored water

in the unit 1 it is necessary to consider the quality of contaminated water. Water began to separate from the solid phase of the accumulated wastes about two years after the launching in operation, i.e.early in 1982. With respect to the character of wastes the values of sulphate were increasing regularly, the values of other parameters increasing also, though not regularly. The presented concentrations of parameters in the accumulated water of the first unit showed that the separated water must be treated separately. The identified increase from the small number of analyses could not be derived explicitly from the eventual leaching from the unit with respect to the distance of the well from the unit, with regard to the permeability coefficient, disturbances of geological layers, and variable effect of precipitation.

The chemization of the adjacent agricultural land, from where leachates of nitrates and potassium may **be expected to form** a significant component in the chemical contamination of waters, was demonstrated in three wells.

Due to a high content of sulphates and sodium in the water of K 1 it may be assumed that water in wells would be affected by those components. Some heavy metals were also investigated /in $mg.l^{-1}$/. Water from wells prior to unloading wastes contained zinc 0,028 and cadmium 0,007; in the course of operation the values were 0,099 and 0,071, and the water in K 1 showed 0,18 and 0,038. Although certain increase was observed the values determined in water from wells do not attain the permissible concentrations for drinking water.

5. CONCLUSION

The poor **homogeneity** of the geological subsoil of the disposal site induced a varied chemism of groundwater and non-uniform spreading of leaching contaminants from the unit into wells. Changes of volume concentrations are influenced by sand lenses in the rather impervious subsoil. Thus, dilution of contamination may occur due to leaching of precipitation, resulting in the non-uniformity of the trend of concentration variation in time and space. It is to be emphasized that the studied territory does not belong to the areas protected for drinking water supply purpose.

6. REFERENCES

1. Appel,D., Kreusch,J.(1986).Müll u.Abfall 18, 9, 348-357.
2. Christensen,T.H.,Kjeldsen,P.(1984).In ISWA Congress, Proceedings Philadelphia, Sept. 15-20.
3. Herbell,J,D.,Tramovsky,B.(1985). In 8.Mülltechn.Sem.No 58. Techn.Univ. München.

ADSORPTIVE REMOVAL AND RECOVERY OF HALOCARBONS FROM LANDFILL GAS

M. SCHÄFER, H.J. SCHRÖTER

BERGBAU-FORSCHUNG GMBH, D 4300 ESSEN, FRG

1. Introduction

Large amounts of halogenated hydrocarbons (HC) are used in branches of industry and households as solving, cleaning, cooling and extracting agents. A large portion of these compounds end up - via the normal refuse disposal - in landfill sites. Compounds like vinylchloride and dichloroethene probably are produced by biological / chemical reactions in the waste. Since landfill gas contains 50 to 70 % by volume of methane, 22 plants for energy production from landfill gas are run in the Federal Republic of Germany /1/. The technical utilization of landfill gas for combined electricity and heat production systems, however, is handicaped by halogenated hydrocarbons. Many HCs are toxic and **carcinogenic**, and are difficult to degrade biologically. Environmental and legal considerations, as well as corrosive properties, make it necessary to remove HC from the landfill gases prior to utilization /2/.

2. Adsorptive removal of halogenated hydrocarbons from landfill gases

A typical application of activated carbon is the removal and recovery of solvents from waste gases. Thus activated carbon can also be used for removal of halogenated hydrocarbons from landfill gases. At Bergbau-Forschung GmbH an adsorption process for removal and recovery of HCs from landfill gases is under development to reduce HC concentrations of approximately 800 mg/m^3 to values lower than 25.0 mg/m^3.

3. Experimental

For process design, the adsorption behaviour of HCs in single- and multiple-component mixtures towards various activated carbons was determined.

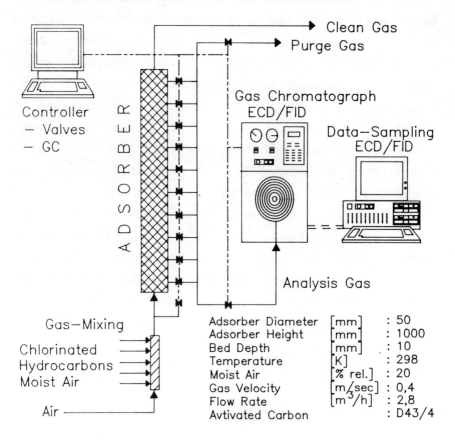

Fig. 1: Adsorption set-up

The raw gas (relative humidity 20 %) was passed through the adsorber filled with activated carbon, and analyzed by gas chromatography at different depth of the adsorber (Fig.1).

4. Test results and discussion

Figure 2 illustrates a typical adsorption diagram. Initially an HC-free gas is analyzed at the first measuring point of the Adsorber after 0.1 m bed depth. After a certain time the outlet concentraion, C, rises to the inlet concentration value of C_0. This adsorption front migrates through the adsorber with increasing test time.

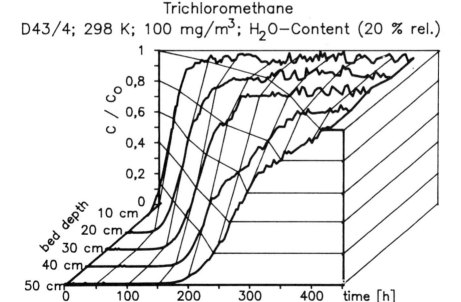

Fig. 2: Breakthrough curves of Trichloromethane

Table 1 shows the results of adsorption experiments with halogenated hydrocarbons after a bed depth of 50 cm. The breakthrough of a compound was reached when the concentration in the analysis gas amounted to 10 % of the C_0 concentration of this compound. The breakthrough times of the different HCs vary in a wide range.

Table 1 : HCs breakthrough times

Compound	breakthrough time (h)
Chloromethane	0,5
Dichloromethane	27,5
Trichloromethane	225,0
Tetrachloromethane	360,0
Chloroethene	1,0
Trichloroethene	270,0
Tetrachloroethene	510,5
Dichlorodifluoromethane	4,0
Trichlorofluoromethane	59,0

Multiple-component adsorption tests resulted in shorter breakthrough times, in comparison to single-component adsorptions. Components with strong interactive forces displaced those which were adsorbed less strongly. Similar results were also found with different humidities in the model gas.

For the description of the adsorption equilibrium and its subsequent calculation various models exist. For microporous adsorbents, such as activated carbons, the Dubinin equation is applicable /3,4/ (I):

$$V = V_s \exp\left[-(\epsilon/\epsilon_0 * \beta)^2\right] \quad (I)$$

V_s = adsorption pore volume
ϵ = adsorption potential
ϵ_0 = characteristic adsorption energy
β = affinity coefficient

Table 2 shows a comparison of some measured and calculated activated carbon capacity values.

Table 2 Compound	activated carbon capacity (mg/g) measured	calculated
Dichloromethane	18,0	24,2
Chloroethene	7,2	8,6
Trichloroethene	156,5	195,5
Tetrachloroethene	380,6	361,5
Trichlorofluoromethane	93,8	95,5

5. Literature

/1/ V. Franzius:
Stand der Deponiegas-Verwertung in der BRD
Europäisches Deponiegas-Forum, München, 1986

/2/ H. Dernbach:
Korrosionsschäden an Gasmotoren in Deponiegasnutzungsanlagen aufgrund CKWs
Recycling International/Jahrbuch 1984

/3/ M.M Dubinin und L.V. Radushkevich
Dokl. Akad. nauk SSSR 55 (1947) 331

/4/ H. Seewald
Untersuchungen zur Sorption organischer Stoffe an Aktivkohle Bergbau-Forschung GmbH, Essen (1974), Dissertation

LANDFILL GAS - ENVIRONMENT PROTECTION AND ENERGY RECOVERY BY A 13.5 MW TOTAL-ENERGY-STATION

J. Schneider,

DEPOGAS- Gesellschaft zur Gewinnung und Verwertung von Deponiegas mbH
Glienicker Straße 100, D-1000 Berlin 39

1. SUMMARY

Berlin's biggest municipal waste disposal site is located in Berlin-Wannsee. From 1954 to 1980 more than 11 million tons of household waste were accumulated there. The site covers an area of about 500,000 squaremeters. A former gravel pit has been refilled up to 40 meters above the surrounding ground. The surface of the site is completely covered with a layer of soil. Decomposition gas, a mixture of Methane (58 %), Carbondioxide (40 %), and more then four hundred different compounds, is produced in the hill as a product of the chemical and bacteriological decomposition of the organic part of the waste. This combustible gas moves through the surface of the site and into the surrounding soil, causing problems regarding the recultivation of the area, and public safety. A feasibility study, based on data from a pilot gas withdrawal plant, proposes the recovery of the gas by a withdrawal system and its utilization in a co-generation plant for the production of electicity and heat. Thus the gas flow across the surface will be reduced by about 70 %, and a potential energy source will be used. A co-generation plant, based on internal-combustion engines with a total electric power output of 4,500 kilowatts, and a thermal power output of 6,500 kilowatts has been constructed in 1987/88. To prevent the origin of Dibenzo-Dioxine and -Furane in the plants exhaust system, a gas pretreatment plant is under construction, to reduce the content of chlorinated Hydrocarbons close to zero in the landfill gas.

2. INTRODUCTION
Disposal in big landfill sites is the common way of handling municipal waste. The greater part of the waste is garbage, a more or less organic material. When dumping the waste, bacteria immediately start decomposing the organic substances. In modern landfill operation, the waste is compressed by heavy machines to get a maximum of waste into the limited space. Aerobic bacteria reduce the content of Oxygen in the porous media, and due to the compression, no further air can get into the waste. With the decrease in Oxygen, anaerobic bacteria start decomposition of the organic material. Parallel to the activity of bacteria, different chemical processes take place.

After a period of about one year, the processes reach a stable state in the dumped waste, and a major part of the decomposition products are gaseous ones. This gas is a mixture of methane (58 %), carbondioxide (40 %), hydrogen, nitrogen, and more than four hundred different compounds as impurities.
In our case, the sum of all impurities is less than 0,1 % in weight. Nevertheless, attention has to be paid to the impurities, because some of them are dangerous chemicals, such as Vinylchloride, Hydrogensulphide, Chlorobenzene, etc., and some are corrosive ones.

The usual path taken by the gas is through the surface of the landfill into the atmosphere. Gas production is high in newly dumped waste with a decrease to zero within some decades. This leads to an oxygen-free zone inside of the landfill, which extends close to the surface. If waste dumping starts beneath the surrounding ground level, e.g. when a former pit is refilled, the gas moves into the surrounding soil too. Plant and tree roots cannot stand an oxygen-free environment for a long period of time. Therefore, it is hard to recultivate a former landfill, and a lot of trees die if they grow close to a landfill.

Methane and some of the impurities are combustible gases. The net caloric value of the gas mixture is about 20,000 kilojoule per cubicmeter.

On the one hand, this combustible potential often causes problems regarding landfill operations, and public safety. On the other hand, it could be a source of energy, if it is taken into account, that dangerous compounds could be **originated** in the flue gas, as there is Dibenzo-Dioxin **and Dibenzo-Furan.**

3. WASTE DISPOSAL SITE
   The Berlin-Wannsee municipal waste disposal site was closed in 1980. Starting in 1954 a former gravel pit with an average depth of 10 meters was refilled, and a hill with a max. height of 40 meters was formed. The area of the landfill covers 50 hectares (500,000 squaremeters). The total dumping space amounts to about 11 million cubicmeters. Domestic garbage or comparable waste is up to 65 % of the content.

   The landfill site will become a public recreation area in the near future. Therefore, the landfill has been covered with a layer of at least 1 meter of soil, mostly clay and sand. Recultivation work was begun in 1965 and will be completed within the next few years.

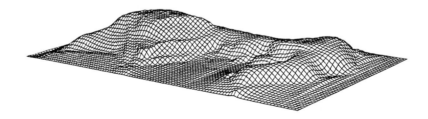

   Figure: The Berlin-Wannsee site from the north. The basic area depicted is 1.560 meters by 710 meters.

4. GAS WITHDRAWAL PILOT PLANT
   A first rough calculation of decomposition gas production in the landfill, done in 1980, showed a volume of about 8,000 cubicmeters per hour. That means a thermal power of about 40 megawatts, based on 50 % of methane in the gas mixture. Therefore, in 1982 a gas withdrawal pilot

plant was constructed, to acquire data about the recoverable gas volume, and gas quality, about the efficiency of the withdrawal system, and about the optimal depth, and distance of the vertical suction wells. Additional research has been done in identifying and quantifying the impurities in the gas mixture and in the flue gas after burning the landfill gas.

Modell-Calculations have been done to assess the time dependent volume of gas production over the next 15 to 20 years, and the recoverable volume of gas.

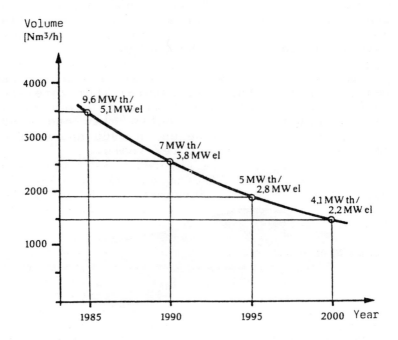

Figure: Time dependent volume of recoverable gas (47 % methane) and the potential co-generation power, regarding internal-combustion engines.

A feasibility study, based on the data from the pilot plant and taking into account the local conditions has shown that the gas utilization in a co-generation plant of about 4,000 kilowatts electric power output will be energetically the most useful and economically the best solution to the problem of gas treatment.

All this work is done mainly by the Hahn-Meitner-Institut Berlin GmbH (HMI), a National Research Institute, which is located close to the landfill (about 500 meter).

5. DECOMPOSITION GAS AS AN ENERGY SOURCE

Parallel to the construction of the pilot plant, a pipeline was constructed from the pilot plant to the heating center of the HMI. A former oil-fired boiler was reconstructed with a burner for landfill gas. Since May 1983 about 65 % of HMIs yearly requirement of heat energy is produced by burning landfill gas from the landfill. That means a saving of about 850,000 liters of oil per year.

The results of the feasibility study, the positive experience with the pilot plant, and the necessary safety of people and vegetation led to the decision to construct a large-scale decomposition gas withdrawal and treatment system.

The Hahn-Meitner-Institut and the local electricity company Berliner Kraft- und Licht (BEWAG)-Aktiengesellschaft made an agreement to construct and to run a large-scale gas recovery and utilization plant. Therefore the DEPOGAS GmbH was founded in September 1986. DEPOGAS started the construction of the gas collection and utilization plant in March 1987.

The decomposition gas is collected by a system of 133 vertical gas suction wells, with single flow-regulation each, to optimize the gas recovery in respect of gas quality and gas flow through the surface. A system of underground pipes connect the suction wells with the compressor station on the landfill site. Three screw-compressors are in operation. After the compression of the gas up to 3.5 bar (abs.), it is chilled by two **chilling-machines** to get the water out before the gas runs into a 1,5 kilometer long pipe to the utilization plant.

A co-generation power plant has been constructed, based on three internal-combustion engines. The total output of the plant is 4,500 kilowatts electric and 6,500 kilowatts thermal power.

The electric power is generated by synchronous generators at 10,000 volts, and fed into the public grid. The thermal power will be used in the buildings of the HMI and in **apartment-houses** in the neighbourhood.

Sciencific work, done on the field of impurities in the landfill gas, as well as in the flue gas from a boiler and a small (150 KW) cumbustion engine has shown the necessity of the careful observation of such impurities.
The origination of Dibenzo-Dioxine **and** -Furane by burning landfill gas led to the decision to construct a gas pretreatment plan, to collect the chlorinated Hydrocarbons from the gas, before it is fueled to the engines.

The pretreatment plant, a counter-stream wash process with a special liquid, two absorption **columns**, two desorption **columns** and a final distillation **column**, where the gas is **purified** to about zero concentration of chlorinated Hydrocarbons, is the first time constructed here.

Regarding the environment protection aims on the landfill site, restricted levels in noise and flue-gas emission of the power plant have been formulated by DEPOGAS. This led e.g. to the necessity of technically reducing the Nitrogenmonoxide and the Nitrogendioxide in the flue gas of the co-generation plant. The so called Selected Catalytic Reduction (SCR) process is the chosen one.
The necessery investment is 22 million Deutsch Marks for the construction of the gas recovery system, gas pretreatment plant and the co-generation plant. The repayment-time for that money will be at least 15 years.

A STUDY OF TRACE COMPONENTS IN LANDFILL GAS FROM THREE UK HOUSEHOLD WASTE LANDFILL SITES

P E Scott, C G Dent, G Baldwin
Waste Research Unit, Harwell Laboratory, Didcot, UK

INTRODUCTION

The Harwell Waste Research Unit (WRU), which is sponsored by the Department of the Environment, has already conducted a number of investigations to identify the compounds present in landfill gas at low concentration. Currently, as part of the Department's extensive ongoing programme of research, investigations have concentrated on the sampling and analysis, particularly for trace components, of gases generated by three household waste landfill sites over an extended period of time. The presence of trace components in landfill gas has implications in terms of both environmental impact and corrosive effects (particularly if the gas is used as a fuel in engines and turbines for electricity generation). A greater understanding of the significance of trace compounds in landfill gas and the major factors which influence their formation and composition is of increasing importance to operators of landfill sites and for landfill gas utilisation schemes.

The sites selected represent a wide range of waste composition, landfill operating conditions and management procedure. At each site landfill gas has been sampled regularly throughout more than three years since the waste was deposited. The scope of this paper allows only a brief overview of results for trace component composition in the samples taken. These are discussed in terms of their known or potential impact on odour, toxicity and corrosivity.

RESEARCH PROGRAMME

Characteristics of the three landfill sites are summarised in Table 1. Patterns of waste degradation have differed markedly at each site. Only that for site B (the deep wet site accepting pulverised waste) has approximated to the classic pattern i.e. early aerobic activity followed by rapid onset of anaerobic conditions and production - once steady state is reached - of a gas containing approximately 60% methane and 40% carbon dioxide(1). By contrast, site A has exhibited an

	Landfill site		
	A	B	C
Description	Shallow, relatively dry	Deep, wet	Deep, relatively dry
Geology	Sandstone and limestone (Upper Jurassic)	Lower coal measures Sandstone (carboniferous)	Limestone with sands and silts (Lower Cretaceous)
Rainfall (mm/yr)	650	950	750
Type of Refuse	Crude	Wet pulverised	Crude
Input Rate (tonnes/d)	50	1000	1000
Method of Operation	Thin-layering	2.5 metre lifts	0.5 metre lifts

TABLE I   Summary of Landfill Characteristics

extended aerobic phase with evidence of only limited and periodic anaerobic decomposition. An extended aerobic phase was also observed at site C, although anaerobic conditions did develop following final restoration and capping.

Details of the sampling and analytical methodologies used in the study are described elsewhere (2,3,4,5). A schematic of the sampling apparatus is provided in Figure 1. The techniques used for analysis of trace components in the samples taken (GC and GC/MS) have been progressively developed and are considered to represent 'state of the art' technology.

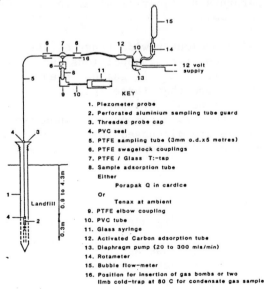

FIGURE 1   Schematic Diagram of Sampling Apparatus

RESULTS

136 trace components have been detected in all, with 109 of these common to all three sites. Total concentrations have varied considerably but have been measured at up to 0.5% of total gaseous emissions. A brief discussion of results is provided below.

Trace Compounds and Odour Impact

Concentrations of individual trace components measured in the gas samples taken at each site have been compared against published odour threshold data. For sites B and C, a large proportion (between 39% and 64%) of these compounds were found at concentrations exceeding odour thresholds. Only in limited cases however were these thresholds exceeded by more than three orders of magnitude. Such compounds were found to be typically the products of anaerobic digestive processes and have tended to belong to the following generic groups: organosulphurs, cyclic compounds, aromatic hydrocarbons, esters and carboxylic acids.

Although the significance of individual compounds has been found to vary with time, a few have been predominant throughout the course of the study. Most noticeable amongst these have been methanethiol and ethyl butanoate. The most odorous emissions have been shown to be associated with the period immediately following waste deposit. During this period, a dilution factor in excess of 1×10^8 may be required to reduce ambient concentrations of the most odorous compounds to below odour thresholds. Monitoring of ambient methane concentrations above a range of landfill sites has shown that such dilution is not always available in the area immediately adjacent to the site. As a consequence, odour problems do arise and these inevitably give rise to complaints. A separate research project by WRU has looked at the ability of a range of different materials to attenuate this landfill gas odour (6). The study has revealed promising results for inexpensive materials such as sawdust and pulverised fuel ash.

Trace Components and Toxicity

Of the total of 136 compounds detected, ten were observed to exceed recognised 8 hour time weighted average toxicity threshold limit values (TLVs) (7,9,10). These compounds were restricted to the following:

a) Organosulphur compounds during the early stages of waste degradation (particularly at sites B and C);
b) Alcohols (again at sites B and C);
c) A number of halocarbons such as dichlorofluoromethane (particularly at sites A and C);
d) Other compounds such as formaldehyde and C_3 benzenes (which only sporadically exceeded their respective TLVs).

The use of individual TLVs is not strictly adequate when considering the toxicity of mixed (in this case gaseous) systems. Of particular concern must be the potential for synergistic toxic effects. However, data on such effects are sparse, in their absence it is recommended that toxic effects be considered additive when they are known to affect the same organ or body function (7) i.e.

$$F = \frac{C_1}{T_1} + \frac{C_2}{T_2} + \frac{C_3}{T_3} + \ldots \frac{C_n}{T_n}$$

where F = TLV factor for target organ or body function.
       Cn = concentration of compound n.
       Tn = individual TLV for compound n.

A toxicity profile for undiluted gas evolved from site B (i.e. the site which approximated most closely to the 'classic' pattern of waste degradation) is shown in Figure 2. This shows the variation in F against time for specific organs or functions of the body. Where values of F exceed unity, the toxic threshold for that particular organ or function has been exceeded.

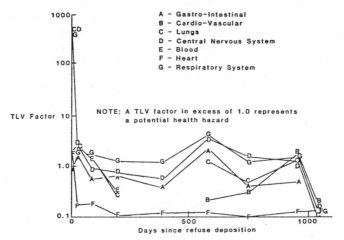

FIGURE 2   Summary of Toxicity Data, Site B

The profile suggests that any potential health risk is associated with the very early period following waste deposit. For this particular site, an early peak value for F of approximately 435 is shown for the lungs, respiratory and central nervous systems. Whilst the atmospheric dilution available to landfill gas should at most sites be sufficient to reduce this value to well below unity, the results nevertheless suggest that it may be advisable under certain circumstances to limit access to areas having received fresh waste for personnel with known respiratory or nervous disorders.

Although eight known or suspected carcinogens have been observed in emissions during the course of this study, discussion of carcinogenic hazards associated with landfill gas is considered beyond the scope of this paper.

Trace Compounds and Corrosion

Landfill gas can be viewed either as a potential environmental hazard or as a potential source of energy. Increasingly in the UK it is being viewed as the latter and approximately 20 landfill gas recovery and utilisation schemes are now in operation. One problem which can arise however (and referred to previously in the introduction) is corrosion of handling and utilisation equipment. This has been attributed not only to the presence of excess moisture (in schemes where the gas has not been adequately dewatered) but also of trace components containing chlorine, fluorine and sulphur.

In West Germany, catastrophic failure of an engine fuelled by landfill gas has been attributed to concentrations of chlorine between 374 mg m^{-3} and 600 mg m^{-3} (8). These concentrations were thought to derive from industrial wastes which were assumed to have been deposited with household waste within the site. Our own studies have also shown very high halocarbon levels (albeit during the very early stage following deposit) in the samples of gas taken from sites A and C. Table 2 shows peak chlorine concentrations for both these sites exceeding those referred to earlier (in addition to which peak fluorine concentrations are also significant). The important point to note is that both sites accept household waste only. The lower concentrations detected at site B (also household waste only) are probably the result of the pulverised nature of the waste (most of the volatile halocarbons being lost during the pulverisation processes) and perhaps also the very wet nature of the site. However, it is

TABLE 2 Total Chlorine and Fluorine in Landfill Gas from Each Site. Total Halocarbons as a Percentage of Total Trace Constituents

SITE	SITE A									SITE B									SITE C								
NUMBER OF DAYS SINCE DEPOSITION OF REFUSE	3	24	87	192	309	583	754	837	957	1040	3	24	87	192	309	583	754	837	957	1040	3	24	87	192	309	583	754	837	957	1040
TOTAL CHLORINE (µg m⁻³)	697.9	300.9	476.5	171.0	348.4	75.8	8.50	0.18	0.48	0.50	72.2	141.5	79.5	50.4	13.1	37.8	37.67	10	180.11	41.78	333.9	571.9	928.2	408.4	311.0	237.0	105.09	129.75	22.25	60.21
TOTAL FLUORINE (µg m⁻³)	57.18	227.72	104.16	32.74	83.27	4.15	1.53	0.029	0.018	0.00	15.0	59.0	18.27	19.20	4.76	13.34	7.37	10	35.89	6.43	82.35	195.72	180.87	28.37	70.74	47.75	26.45	19.03	4.3	4.91
TOTAL HALOCARBONS AS A PERCENTAGE OF TRACE CONSTITUENTS	33.99	37.99	36.15	23.27	33.94	59.03	23.12	3.18	2.13	1.39	2.30	3.52	8.43	8.48	1.49	0.94	4.17	10	14.98	7.72	10.23	29.98	30.49	18.26	34.12	8.99	9.29	10.05	3.20	9.24

interesting that halocarbon concentrations measured in recent samples taken at site B have shown an upward trend; it is suspected that this is a consequence of absorption/desorption reactions occurring within the pulverised waste.

The corrosive potential of the gas generated at each site is discussed in more detail elsewhere (4). The overall conclusion however is that recovery and utilisation equipment (particularly equipment used to generate electricity) may suffer corrosion problems if adequate measures are not taken to monitor the gas for halocarbon concentrations and ensure that adequate protection is provided.

REFERENCES

1) Rees J F (1981)
   Proceedings of Landfill Gas Symposium, Harwell Laboratory, May 1981. Paper 2.

2) Brookes B I and Young P J (1983)
   Talanta, Vol 30, No 9, pp665-676.

3) Young P J and Parker A (1983)
   Waste Management and Research, Vol 1, No 3, pp213-226.

4) Dent C G, Scott P and Baldwin G (1987)
   Proceedings of the Conference on Energy from Landfill Gas, Solihull, October 1986, pp130-150.

5) Mindrop R (1978)
   J. Chromatog. Sci., Vol 16, p380.

6) Emberton J R, Scott P E and Dent C G (1987)
   The Attenuation of Landfill Gas Odour Using Waste
   Materials. Proceedings of 2nd International Conference
   on New Frontiers for Hazardous Waste Management. EPA.
   Pittsburgh, September 1987.

7) American Conference of Governmental and Industrial
   Hygienists (1987)
   Threshold Limit Values and Biological Exposure Indices
   for 1987-88 ACGIH.

8) Dernbach H (1985)
   Waste Management and Research, Vol 3, No 2, pp149-159.

9) Mackinson F W, Stricoff R S and Partridge L J (1980)
   NIOSH/OSHA, Pocket Guide to Chemical Hazards.
   Publication No 78-210.

10) United Kingdom Health and Safety Executive (1986)
    Occupational Exposure Limits, Guidance Notes EH/40/86.
    HMSO.

LEACHATE QUALITY FROM CONTAINMENT LANDFILLS IN WISCONSIN

L. Sridharan, P. Didier, DNR, Madison, Wisconsin, USA

INTRODUCTION

Since 1975, the Wisconsin Solid Waste program has required collection and treatment of leachate from landfills. As a part of this requirement, the Wisconsin landfills with leachate collection systems are required to periodically sample and chemically characterize the leachate quality. Wisconsin currently has leachate quality data from 56 landfills some of which were built in the mid '70's and some were built as recent as 1987.

The summary of the data gathered and presented in this paper is not intended to be the final answer to predicting leachate quality, but hopefully when examined with other available information on the site, a more accurate assessment of a given site's expected leachate chemical characteristics will be possible. This paper is an update of the information presented by Kmet and McGinley in 1982 (1) at the 5th Annual Madison Conference.

METHODS

To initiate this study, all Wisconsin landfills with leachate collection systems were first identified. The Wisconsin Solid Waste personnel familiar with the landfills, selected those sites which are likely to offer representative leachate quality. The landfills with leachate likely to be diluted by the groundwater or the surface water were eliminated from the list. Supplemental data were gathered for the selected landfills from their case files and published information (1, 2). A large number of the selected landfills accept municipal and commercial wastes; some landfills accept demolition wastes; some landfills are built by individual paper mill industries to dispose of their industrial sludges; and some landfills are built by the utilities to dispose of their incinerator ashes. The selected landfills have various design types such as clay liner, natural attenuation and zone of saturation. Their design capacity varies from 50,000 to 15 million cubic yards. For each selected landfill, the available chemical data from the headwell, leachate tank and leachate manhole are included in this study.

The leachate samples were retrieved by the site operator or their consultant, so the exact method of sampling is often not known. Most samples were probably retrieved by dipping a PVC bailer for inorganic parameter sampling and transferred to bottles with preservative for storage and transport to a laboratory for analyses. For organic parameters, a teflon bailer and vials with septum tops for sample collection and storage, respectively, were generally used. Some samples were filtered in the field prior to preservation while others were field preserved and later filtered in the laboratory. Most laboratory operators followed procedures for analyses such as the APHA Standard Methods or the EPA approved methods (3, 4).

RESULTS & DISCUSSION

Table 1 summarizes the analytical results for each parameter tested. The table includes the total number of samples tested, the number of sample values above the detection limit, and the range and the median for values above the detection limit. The wide range in the concentrations may partially be attributed to the variability in the sampling methods and the use of many acceptable alternative analytical methods. No attempt is made in this publication to correlate the design, age and the type of wastes accepted to the leachate quality, although it is expected that all these factors play important roles.

Table 2 offers a comparison between the values of leachate quality and the range and maximum values found in Wisconsin's groundwater. Of the 18 parameters compared, with the exception of nitrate plus nitrite, the median values of all parameters exceeded the range of mean values

TABLE 1: STATISTICAL VALUES FOR PARAMETERS IN LEACHATE

PARAMETER	NO. OF SAMPLES	NO. > D.L.	RANGE (FOR SITES WHERE DETECTED)	MEDIAN	
ACENAPTHENE	26	2	13.9 - 21.3	17.60	UG/L
ACRYLONITRILE	49	1	1000	1000.00	UG/L
ALKALINITY, CARBONATE	1	1	1200	1200.00	MG/L
ALKALINITY, TOTAL	790	786	4.9 - 44400	2300.00	MG/L
ALUMINUM, TOTAL	7	5	1.5 - 900	600.00	UG/L
ANTHRACENE	27	1	16.1	16.10	UG/L
ANTIMONY, TOTAL	26	8	0.08 - 0.56	0.18	UG/L
ARSENIC, DISSOLVED	8	3	7 - 18	10.50	UG/L
ARSENIC, TOTAL	153	83	0.4 - 52000	19.00	UG/L
BARIUM, DISSOLVED	5	5	150 - 2400	375.00	UG/L
BARIUM, TOTAL	70	59	12 - 33300	320.00	UG/L
BENZENE	35	14	1 - 1630	11.10	MG/L
BERYLLIUM, TOTAL	27	1	0.24	0.24	UG/L
BIOCHEMICAL OXYGEN DEMAND	1	1	47	47.00	MG/L
BIS-2-ETHYLHEXYLPHTHALATE	26	5	91 - 7900	1050.00	UG/L
BOD	790	777	1 - 72000	1710.00	MG/L
BORON, DISSOLVED	4	4	4.25 - 6.8	5.70	MG/L
BORON, TOTAL	116	97	0.12 - 133	4.20	MG/L
BROMODICHLOROMETHANE	39	1	2490	2490.00	UG/L
BUTYLBENZYLPHTHALATE	27	2	10 - 64.1	37.05	UG/L
CADMIUM, DISSOLVED	13	4	0.5 - 30	17.00	UG/L
CADMIUM, TOTAL	138	67	0.8 - 200	20.00	UG/L
CALCIUM, DISSOLVED	3	3	80 - 380	317.00	MG/L
CALCIUM, TOTAL	78	78	25 - 2100	354.00	MG/L
CARBON TETRACHLORIDE	50	3	3 - 995	28.00	UG/L
CARBON, TOTAL ORGANIC	227	226	0.3 - 45100	1438.75	MG/L
CHLORIDE	893	891	0.89 - 27100	486.00	MG/L
CHLOROBENZENE	50	7	3 - 188	25.20	UG/L
CHLORODIBROMOMETHANE	36	1	31	31.00	UG/L
CHLOROETHANE	50	11	2 - 730	17.00	UG/L
CHLOROFORM	50	7	4.4 - 16	7.14	UG/L
CHROMIUM, DISSOLVED	8	5	2.93 - 5700	69.00	UG/L
CHROMIUM, HEXAVALENT	7	1	70	70.00	UG/L
CHROMIUM, TOTAL	165	100	2 - 2800	100.00	UG/L
CHRYSENE	27	1	19.4	19.40	UG/L
CIS-1,3-DICHLOROPROPENE	40	1	2.5	2.50	UG/L
COBALT, TOTAL	1	1	3.402	3.40	MG/L
COD	956	949	1.2797 - 79200	1823.00	MG/L
COPPER, DISSOLVED	14	9	20 - 100	50.00	UG/L
COPPER, TOTAL	113	76	7 - 35000	50.00	UG/L
CYANIDE, TOTAL	91	63	0.004 - 555	0.08	MG/L
DI-N-BUTYL PHTHALATE	26	4	13 - 540	28.70	UG/L
DI-N-OCTYL PHTHALATE	26	5	16.1 - 542	110.00	UG/L
DIBROMOCHLOROMETHANE	24	2	22 - 160	91.00	UG/L
DICHLORODIFLUOROMETHANE	37	2	100 - 242.1	171.05	UG/L
DICHLOROMETHANE	41	25	27.6 - 58200	483.00	UG/L
DIETHYL PHTHALATE	26	11	12 - 230	44.00	UG/L
DIMETHYL PHTHALATE	26	1	66	66.00	UG/L
DISSOLVED SOLIDS, TOTAL	172	172	6.36 - 50380	1396.00	MG/L
ETHYLBENZENE	52	30	1 - 1680	43.50	UG/L
FLUORANTHENE	27	4	9.56 - 723	39.10	UG/L
FLUORENE	27	2	21 - 32.6	26.80	UG/L
FLUORIDE, DISSOLVED	3	2	0.2 - 0.76	0.48	MG/L
FLUORIDE, TOTAL	41	37	0.1 - 166	0.36	MG/L
FLUOROTRICHLOROMETHANE	50	7	1 - 183	34.00	UG/L
FORMALDEHYDE	2	2	1 - 1.4	1.20	MG/L
HALOGEN, TOTAL ORGANIC	73	72	0.0039 - 33400	623.50	UG/L
HARDNESS, CALCIUM	1	1	2440	2440.00	MG/L
HARDNESS, MAGNESIUM	1	1	1720	1720.00	MG/L
HARDNESS, TOTAL	1000	999	1.14 - 40000	2000.00	MG/L

TABLE 1: STATISTICAL VALUES FOR PARAMETERS IN LEACHATE

Parameter				
IRON, DISSOLVED	385	378	0.015 - 1580	25.30 MG/L
IRON, TOTAL	518	518	0.024 - 57300	73.50 MG/L
ISOPHORONE	26	13	3.18 - 520	76.00 UG/L
ISOPROPYLBENZENE	11	1	1	1.00 UG/L
LEAD, DISSOLVED	14	10	1.6 - 300	72.50 UG/L
LEAD, TOTAL	138	99	1.4 - 1240	100.00 UG/L
LITHIUM, DISSOLVED	3	3	1.37 - 820	450.00 UG/L
MAGNESIUM, DISSOLVED	3	3	210 - 275	230.00 MG/L
MAGNESIUM, TOTAL	56	56	6.55 - 780	91.00 MG/L
MANGANESE, DISSOLVED	100	98	3.87 - 60200	835.00 UG/L
MANGANESE, TOTAL	93	93	1.03 - 81000	1310.00 UG/L
MERCURY, DISSOLVED	8	3	3 - 500	5.00 UG/L
MERCURY, TOTAL	135	25	0.1 - 16	1.00 UG/L
METHYLETHYLKETONE	11	2	2100 - 37000	19550 UG/L
MOLYBDENUM, TOTAL	15	9	0.005 - 7.7	0.21 MG/L
NAPTHALENE	21	10	4.6 - 186	33.75 UG/L
NICKEL, DISSOLVED	2	2	100 - 180	140.00 UG/L
NICKEL, TOTAL	101	92	0.18 - 60000	260.00 UG/L
NITRATE NITROGEN, DISS	1	1	0.1	0.10 MG/L
NITRATE NITROGEN, TOTAL	97	52	0.1 - 240	2.05 MG/L
NITRITE NITROGEN, TOTAL	33	13	0.005 - 1.46	0.06 MG/L
NITRITE PLUS NITRATE, TOT	79	61	0.015 - 806	1.02 MG/L
NITROGEN, AMMONIA, TOTAL	152	151	0.13 - 5810	217.00 MG/L
NITROGEN, INORGANIC, TOTAL	1	1	93.4	93.40 MG/L
NITROGEN, KJELDAHL, DISS	18	18	0.57 - 806	72.25 MG/L
NITROGEN, KJELDAHL, TOTAL	280	280	0.5 - 3600	237.85 MG/L
NITROGEN, ORGANIC, TOTAL	2	2	3 - 6	4.50 MG/L
P-CHLORO-M-CRESOL	22	1	25	25.00 UG/L
P-DICHLOROBENZENE	37	12	2 - 250	14.00 UG/L
PCBS	11	1	0.0036	0.00 MG/L
PENTACHLOROPHENOL	25	1	25	25.00 UG/L
PH-FIELD	1263	1263	3.35 - 10.1	6.70 SU
PH, LAB	26	26	5.7 - 8.6	6.83 SU
PHENANTHRENE	27	5	8.1 - 1220	50.70 UG/L
PHENOL	27	16	1.1 - 2170	174.00 UG/L
PHENOLICS, TOTAL	128	120	0.052 - 19000	619.00 UG/L
PHOSPHORUS, TOTAL	209	208	0.005 - 240	1.81 MG/L
POTASSIUM, DISSOLVED	3	3	210 - 410	324.00 MG/L
POTASSIUM, TOTAL	77	77	2.74 - 1425	100.00 MG/L
SELENIUM, TOTAL	103	28	1 - 1020	14.90 UG/L
SILICA, TOTAL	1	1	34	34.00 MG/L
SILVER, TOTAL	86	24	0.0019 - 120	0.04 UG/L
SODIUM, DISSOLVED	4	4	18 - 1360	745.00 MG/L
SODIUM, TOTAL	252	251	3.95 - 3370	360.00 MG/L
SOLIDS, TOTAL	6	6	1668 - 22156	9180.00 MG/L
SPECIFIC CONDUCTANCE-FIELD	1133	1132	7.1 - 72500	5110.00 MICR
SPECIFIC CONDUCTANCE, LAB	11	11	300 - 19770	3540.00 MICR
STYRENE	11	1	2	2.00 UG/L
SULFATE, DISSOLVED	2	2	121 - 638	379.50 MG/L
SULFATE, TOTAL	645	573	0.01 - 4150	90.00 MG/L
SULFIDE, TOTAL	37	15	1 - 125	1.45 MG/L
SULFITE	3	2	1 - 20	10.50 MG/L
SUSPENDED SOLIDS, TOTAL	601	600	1 - 89800	129.00 MG/L
TANNIN AND LIGNIN, COMBINED	52	48	0.12 - 264	1.94 MG/L
TEMPERATURE, WATER	55	55	2 - 122	12.00 CENT
TETRACHLOROETHYLENE	52	10	1 - 232	16.30 UG/L
TETRAHYDROFURAN	12	6	410 - 1400	730.00 UG/L
THALLIUM, TOTAL	26	6	28 - 930	880.00 UG/L
TIN, DISSOLVED	1	1	6500	6500.00 MG/L
TOLUENE	53	42	1 - 11800	360.00 UG/L
TOLUENE BY GC-MS	16	3	18 - 110	63.70 MG/L
TRANS-1,3-DICHLOROPROPENE	38	1	2.5	2.50 UG/L
TRIBROMOMETHANE	50	1	47	47.00 UG/L
TRICHLOROETHYLENE	53	12	1 - 372.2	19.00 UG/L
VINYL CHLORIDE	42	12	10 - 3000	230.00 UG/L
XYLENE	13	10	9.4 - 240	72.50 UG/L
ZINC, DISSOLVED	18	17	30 - 2000	210.00 UG/L
ZINC, TOTAL	115	108	3.6 - 71000	889.50 UG/L

found in the Wisconsin groundwater. The median values for 11 parameters in leachate exceeded the maximum groundwater concentrations.

Another useful benchmark in obtaining a perspective on the contaminant concentrations in leachate is the U.S. drinking water standards and the Wisconsin Groundwater Enforcement Standards. The data show (Table 3) a trend similar to the groundwater quality data in Table 2. Almost 50 percent of the median values for the chemical parameters in leachate exceeded the drinking water standards and Wisconsin Groundwater Enforcement Standards.

This analysis confirms that the leachates from solid waste landfills have chemical strengths significantly greater than the background groundwater quality . The results emphasize the importance of a well designed leachate collection system to intercept the leachate at the base of the landfill. The study also underscores the importance of a well constructed liner for the landfill to minimize the percolation of the higher strength leachate in to the groundwater.

1) Kmet, P. and McGinley, P. M. (1982). Chemical characteristics of leachate from municipal solid waste landfills in Wisconsin, Fifth Annual Madison Conference of Applied Research and Practice on Municipal and Industrial Wastes, Madison, WI.

2) Kmet, P. Mitchell, G. and Gordon, M. (1986). Leachate Collection Systems and Performance - Wisconsin's Experience, Ninth Annual Madison Waste Conference, Madison, WI.

3) American Public Health Association (1976). Standard Methods for the Examination of Water and Wastewater, APHA NY. NY

4) U.S. EPA (1977). Test Methods for Evaluating Solid Waste. Laboratory Manual SW 846, Volumes I & II.

5) Hindall, S. M. (1979). Groundwater Quality in Selected Areas in Wisconsin. Water Resources Investigation Open File Report, U. W. Department of Interior Geological Survey, 79-1594.

TABLE 2: COMPARISON OF WISCONSIN GROUNDWATER AND LEACHATE

PARAMETER	IN WISCONSIN GROUNDWATER (5) RANGE OF MEAN	MAXIMUM	IN WISCONSIN LEACHATE RANGE	MEDIAN
ALKALINITY	120 - 250	500	4.9 - 44400	2300
AMMONIA-N	ND - 0.01	3.2	.13 - 5810	217
BORON	0.02- 0.07	.65	.12 - 133	4.2
CALCIUM	40 - 140	875	25 - 2100	354
CHLORIDE	5 - 50	3510	0.89- 27100	486
HARDNESS	100 - 540	2875	1.14- 40000	2000
IRON	0.05- 0.55	46	.014- 57300	73.5
MAGNESIUM	15 - 45	169	6.55- 780	91
MANGANESE	0.04- 0.075	1.4	1.03- 81000	1310
NO2+NO3	1 - 11	30	.015- 806	1.02
PH	6 - 8.5		3.35- 10.1	6.7
POTASSUIM	1 - 5	42	2.74- 1425	100
SODIUM	10 - 50	1830	3.95- 3370	360
SPEC. COND	100 - 500	15000	7.1 - 72500	5110
SULFATE	10 - 275	2000	0.01- 4150	90
TDS	200 - 650	8420	6.36- 50380	1396
TKN	0.01- 0.3	3.4	0.5 - 3600	237
TOC	1 - 6	37	0.3 - 45100	1438

ALL CONCENTRATIONS IN MG/L EXCEPT PH (STD UNITS)
AND SP. COND (UMHOS/CM)

TABLE 3: COMPARISON OF LEACHATE CONCENTRATIONS TO DRINKING WATER STANDARDS

PARAMETER	NO. OF SAMPLES	RANGE	MEDIAN	DRINKING WATER STANDARD
ARSENIC, TOTAL	153	0.4 - 52000	10.00	50.00 UG/L
BARIUM, TOTAL	70	10 - 33300	272.50	1000.00 UG/L
BENZENE	35	1 - 10000	10.00	0.67 UG/L *
CADMIUM, TOTAL	138	0.2 - 4000	10.00	10.00 UG/L
CHLORIDE	893	0.89 - 27100	486.00	250.00 MG/L
CHROMIUM, TOTAL	165	1 - 20000	50.00	50.00 UG/L
COPPER, TOTAL	113	3 - 35000	50.00	1.00 MG/L
CYANIDE, TOTAL	91	0.004 - 555	0.04	0.46 MG/L *
DICHLOROMETHANE	41	1.32 - 58200	100.00	150.00 UG/L *
DISSOLVED SOLIDS, TOTAL	172	6.36 - 50380	1396.00	500.00 MG/L
ENDRIN	26	0.4 - 200	10.00	0.20 UG/L
FLUORIDE, TOTAL	41	0.1 - 166	0.33	2.20 MG/L
IRON, TOTAL	518	0.024 - 57300	73.50	0.30 MG/L
LEAD, TOTAL	138	1 - 1240	63.00	50.00 UG/L
MANGANESE, TOTAL	93	1.03 - 81000	1310.00	0.05 MG/L
MERCURY, TOTAL	134	0.1 - 1000	0.60	2.00 UG/L
P-DICHLOROBENZENE	37	0.5 - 250	10.00	750.00 UG/L *
SELENIUM, TOTAL	103	0.7 - 5000	5.00	10.00 UG/L
SILVER, TOTAL	86	0.0007 - 120	2.00	50.00 UG/L
SULFATE, TOTAL	645	0.01 - 4150	73.00	250.00 MG/L
TETRACHLOROETHYLENE	52	0.53 - 10000	16.30	1.00 UG/L *
TOLUENE	53	1 - 11800	240.00	343.00 UG/L *
TOXAPHENE	26	0.5 - 200	10.00	0.00 UG/L *
TRICHLOROETHYLENE	53	1 - 10000	19.00	1.80 UG/L *
VINYL CHLORIDE	42	0.18 - 10000	75.50	0.01 UG/L *
XYLENE	13	2 - 240	64.00	620.00 UG/L *
ZINC, TOTAL	115	3.6 - 71000	650.00	5000.00 UG/L
1,1-DICHLOROETHYLENE	44	1 - 400	2.80	0.24 UG/L *
1,1,1-TRICHLOROETHANE	53	1 - 10000	38.00	200.00 UG/L *
1,1,2-TRICHLOROETHANE	53	1.5 - 10000	10.00	0.60 UG/L *
1,2-DICHLOROETHANE	52	1 - 10000	10.00	0.50 UG/L *

* - STATE OF WISCONSIN ENFORCEMENT STANDARD

ADVANCED LANDFILL TECHNOLOGY FOR JOINT LEACHATE AND GAS MANAGEMENT

Philip P. Stecker, P.E., Donohue & Associates, Inc.
4738 North 40th Street, Sheboygan, Wisconsin 53083 U.S.A.

INTRODUCTION

Advanced technology has been developed for the 3.8 million cubic meter Outagamie County, Wisconsin, U.S.A., landfill, including design from the beginning for joint leachate management/energy recovery and convenient internal television camera inspection. The East Landfill Expansion, which opened in January, 1987, receives 900 metric tons per day of solid waste and has a projected 12-year service life.

The landfill expansion was designed with substantially more piping system access locations, leachate/landfill gas collection piping, and granular material for gas collection compared to conventional landfills. These features were possible, as well as cost effective, because of their installation at the time of construction -- as part of multipurpose systems. The innovations were given a National Honor Award for Engineering Excellence in 1987 by the American Consulting Engineers Council of Washington, D.C.

Where possible, conventional landfill or sewer system design features were used for the overall site, piping network, and structures. New engineering concepts were developed where necessary. Various innovative, integrated, multipurpose, and redundant systems are included.

LANDFILL BASE CONFIGURATION

The landfill base includes a 1.5-meter thick recompacted clay liner with a relatively steep 3 percent slope and a maximum leachate flow distance of approximately 30 meters from the clay liner high point to the low point at the leachate collection pipeline. The ridge-and-trough configuration segments the base into 22 cells, each containing a separate leachate collection pipeline.

A 5,100-meter network of piping is provided at the base of the landfill for leachate and gas collection. The majority of the system consists of 15-centimeter diameter Schedule 80 PVC piping with bottom perforations for leachate entry and top perforations for landfill gas entry. The perforated lateral pipes are interconnected along the landfill perimeter with 20-centimeter diameter nonperforated piping. A precast concrete manhole is located at each of the pipeline junctions. Leachate flows to two low points in the landfill where pump stations are used to remove leachate.

PIPING SYSTEM MANHOLES

Custom design features were developed for the manholes within the landfill. Manholes were incorporated at strategic locations in the piping system. The manholes were included because independent leachate pipeline cleaning activities performed at older portions of the landfill demonstrated that manholes can serve an important purpose as entryways into the landfill for monitoring and maintenance.

Manholes inside of a landfill should not be used for routine personnel entry because of the potential dangers associated with landfill gas, leachate, and differential settlement. Manholes can be beneficial, however, for seeing to the bottom of the landfill. By working from the top of a manhole, operations personnel can visually determine whether any liquid surcharging exists at the bottom of the site. Such observations can determine whether pipeline blockages may exist in any areas.

Another purpose for a manhole would be to serve as the downstream starting point for performing jet cleaning of pipelines. Without personnel entry into the manhole and with the dedication of sufficient effort, it is sometimes possible to lower a jet cleaning nozzle and hose from the top and insert the cleaning device into the pipeline at the bottom of the manhole.

Manholes can also serve as backup locations for leachate removal if a pipeline blockage prevents flow to the pump station. Having manholes at critical locations within the piping network could be beneficial for leachate removal.

During refuse filling, manholes within a landfill can often be misaligned or damaged. To reduce the effects of refuse placement and compaction on manhole integrity, tie rods are provided to hold sections of manhole together. Four vertical tie rods are provided around each manhole. Steel brackets at the midpoint of each manhole section fasten the tie rods to the manhole wall. The tie rods are used to draw manhole sections together and maintain straight alignment and tight-fitting joints. Joint sealant is used between manhole sections to prevent undesired landfill gas, air, or leachate migration. Each manhole is closed at the top with a sealed, hinged lid to prevent landfill gas escape and air entry into the system. Without maintaining sealed conditions for the manholes and other risers, air could be drawn into the leachate/gas collection piping. Gas monitoring must be performed to ensure that sealed conditions are maintained.

PIPING SYSTEM ACCESS RISERS

Sixty-five access risers are provided at critical locations throughout the piping network for long-term inspection and maintenance. Included are manhole, sideslope, and center access risers. Long-radius bends are provided for all of the risers. With these risers, convenient access is provided to the 5.2-kilometer piping system for cleaning equipment, monitoring devices, and television cameras.

Access risers are included at 4 of the 22 manhole locations. The risers connect at the bottom to the horizontal leachate/gas collection pipelines. The risers are attached to the side of the manholes and are used for routine cleaning of the 20-centimeter diameter nonperforated transmission pipes. The four manhole locations were selected to keep the maximum pipeline cleaning length less than 250 meters. Jet cleaning can be accomplished from the downstream end of each pipe segment.

The 15-centimeter diameter perforated pipelines cross the landfill base at a 60-meter spacing. Jet cleaning of these lines is accomplished from sideslope risers located at 22 places in the landfill. The risers are installed on 3:1 sideslopes of the landfill. The bottom of each sideslope riser connects to a manhole.

A 7.6-meter radius specially fabricated PVC long-sweep bend creates the transition between the sideslope riser and a horizontal alignment at the bottom of the manhole. Each sideslope riser connects at the low point of a perforated pipeline and thereby allows for cleaning from the downstream end and pulling of any debris to the manhole and leachate transmission pipeline.

Vertical center access risers are provided at the midpoint of each perforated pipeline in the center of the landfill. Eleven center access riser systems are included. Concrete encasement is provided at the bottom of each center riser assembly to hold the piping together and create a secure foundation for the structure.

COMPUTER-AIDED DEVELOPMENT

The landfill design, drafting, and construction was accomplished with computer-aided techniques. Computerized methods were used to develop the earthwork, contouring, pipeline, structural, and mechanical design features. Numerous landfill design options can be evaluated with computerized methods, a process too costly and time-consuming with manual methods. During construction, detailed geometric information was provided using a computer data management system.

LEACHING OF METALS FROM DESULPHURIZATION WASTES

Margareta Wahlström[1]) & Jussi Ranta[2])
[1]) Chemical Laboratory
[2]) Laboratory of Fuel Processing and Lubrication Technology
Technical Research Centre of Finland, SF-02150 Espoo, Finland

1. Introduction

Environmental impacts of desulphurization wastes are primarily due to materials leached from the wastes when used in soil construction works or disposed on landfills. The desulphurization wastes contain minor amounts of substances hazardous to the environment, primarily metals. However, it is not possible to estimate the amount of components leaching from the wastes at a sufficient accuracy only on the basis of composition data.

Substances leached from the waste of a wet-dry method and from the ash from fluidized-bed combustion are discussed in the following. The effect of granulating the waste of the wet-dry method on the leaching of metals was also studied.

2. Materials

The materials included in the study are as follows:

* waste from a wet-dry process (Fläkt process),
* ash from fluidized-bed combustion (Pyroflow process)
* fly ash from entrained-bed combustion of coal,
* reference materials: a piece from a concrete well ring, crush made from blast furnace slag, expanded clay.

The composition of the waste of the wet-dry method, the ash from fluidized-bed combustion, and the coal fly ash are given in Tables 1 and 2. The contents of hazardous metals are typically less than 100 ppm. The metal content of the waste of the wet-dry method is usually lower than that of the ash from fluidized-bed combustion or of the coal fly ash. The chemical composition of the reference materials was not determined.

Table 1a. Chemical composition of the waste of the wet-dry method.

$CaSO_3 \times 0,5\ H_2O$	47 %
$CaSO_4 \times 2\ H_2O$	18 %
$Ca(OH)_2 + CaCO_3$	appr. 27 %
$CaCl_2 \times 4\ H_2O$	0,5 %
Fly ash content	8 %

Table 1b. Chemical composition of the ash from fluidized-bed combustion and of the coal fly ash.

	Ash from fluidized-bed combustion	Coal fly ash
CaO (total [1])	10,8 %	3,6 %
$CaSO_4 \times 2H_2O$	17,2 %	
$CaSO_3 \times 0,5\ H_2O$	<1 %	
$CaCO_3$	0,3 %	
$CaCl_2 \times 4H_2O$	0,08 %	
CaO	5,0 %	
SiO_2	37,3 %	51,4 %
Al_2O_3	15,3 %	14,2 %
Fe_2O_3	5,0 %	7,6 %
K_2O	0,9 %	2,1 %
Na_2O	0,6 %	1,1 %
MgO	1,4 %	1,7 %
TiO_2	7,2 %	0,6 %
Ignition residue	81 %	97 %

[1]) all calcium is assumed to occur as oxides

Table 2. Contents of hazardous metals in desulphurization wastes and coal fly ash.

Metal		Waste of wet-dry process	Ash from fluid. bed combustion	Coal fly ash
As	ppm	10	21	21
Ba		190	750	1600
Cd		<5	<5	<5
Co		10	17	39
Cr		<80	110	170
Mo		30	11	30
Ni		<50	63	110
Pb		15	<1	80
Se		4	6	5
V		14	130	140
Zn		-	40	-

3. Leaching tests

The substances leached from desulphurization wastes were determined with the aid of shaking and column tests. In the shaking test, the sample and water (pH 4) are shaken for about 24 hours and the mixture is then filtered. In the column test, and about 20 cm thick waste layer is sprinkled with water (pH 4) from above. The composition of the filtrate obtained in the leaching tests is dependent on the total water amount contacted a certain waste amount. The quality of the filtrate water is described as a function of the water-waste ratio (L/S).

The shake and column tests were carried out for a mixture of waste of the wet-dry method (FGD) and coal fly ash (FA) their ratio being FGD:FA = 65:35. Water was added to the mixture in order to initiate the puzzolanic reactions. The waste mixture was stored about one month before conducting the shake test and 3 days before starting the column test. Substances leached from the coal fly ash used in the mixture were also determined with the shake test.

Granulates were made from the mixture of waste and fly ash on a disc granulator. The FGD:FA ratio was 40:60. The granulates were screened using a 10 mm sieve. The shaking test was made for the sieved granulates.

The substances leached from the fluidized-bed ash were determined with the shaking test and the column test. Water was added to the ash and the mixture was stored about one month before the shake test and 3 days before the column test.

The substances leached from the reference materials were determined with the shake test. The concrete piece taken from the well ring was crushed with a jaw-cone crusher. The crushed sample was screened on a 2 mm sieve. The test was carried out for the sieved material.

Part of the results obtained in the leaching tests are presented in Table 3. Unfortunately, the handling of the column tests with the fluidized-bed ash in not yet completed.

Table 3. Results of the leaching tests.

	Coal fly ash	Waste of wet-dry process				Fluid. bed ash	Reference materials		
		Granulate (40% FGD + 60% FA)	Mixture (65% FGD + 35% FA) Shaking test	Column test 0–0,5	Column test 0,5–0,85		Well ring	Blast-furnace slag Crush	Expanded clay
L/S	1,2	1	1,6			1	1	1	1
pH	12,6	10,5	12,6	9,7–12,4	8,2–12,0	10,2	12,7	12,0	9,5
NO$_2^-$ – mg/l	<0,5	–	590	1400	390	–	–	–	–
NO$_3^-$	<0,5	–	240	520	110	–	–	–	–
Cl$^-$	–	–	–	2100	880	–	–	–	–
As	0,011	0,002	0,022	0,008	<0,002	0,018	0,002	<0,002	0,11
Ba	0,88	0,35	1,69	0,50	0,30	0,2	0,58	0,23	0,11
Cd	0,0008	<0,0001	0,050	0,0009	<0,0005	<0,0005	<0,0005	<0,0005	<0,0005
Cr	1,49	1,92	0,28	1,66	0,30	0,010	0,035	0,001	0,006
Cu	0,009	0,048	0,008	0,023	0,03	0,002	0,014	0,002	<0,01
Mo	3,04	1,46	0,99	1,35	1,0	0,14	0,012	0,008	0,11
Ni	0,004	0,002	0,008	0,008	<0,01	<0,001	0,0003	0,0002	<0,01
Pb	0,007	<0,0005	0,018	0,014	0,002	0,005	0,008	0,004	0,002
Se	0,097	0,021	0,045	0,050	<0,01	–	0,004	0,001	<0,01
Zn	0,032	0,003	0,015	0,23	0,05	0,003	0,013	0,013	0,02

4. Discussion

The following conclusions can be drawn from the results of analysis obtained so far:

1. The water filtrates of the mixture of the waste of the wet-dry method and fly ash, as well as those of the granulates, coal fly ash and fluidized-bed ash were alkaline. The pH of the filtrate of fluidized-bed ash was slightly lower than that of the filtrate of the waste of the wet-dry method.
2. The results of the shaking tests and the column tests correlate with each other.
3. The amount of metals leached from the fluidized-bed ash is lower than that leached from the waste mixtures of the wet-dry method.
4. The metal contents of the filtrates made from the waste of the wet-dry method increase when raising the proportion of fly ash.
5. The granulation of desulphurization waste did not reduce leaching of substances.
6. The amount of chrome and molybdenum leached from the waste of the wet-dry process is higher than that from the reference materials. No significant difference was found in the contents of the other studied materials between the desulphurization wastes and the reference materials. The amount of arsenium leached from the expanded clay sample was even higher than that leached from the desulphurization wastes.

Removal of chlorinated hydrocarbons from landfill gas

M. Werner, G. Olderdissen

1. Introduction

Utilization of gas from landfills has within the last 10 years been developed so that 26 plants for generation of electricity and heat only in the Federal Republic of Germany are established, Franzius (1987). In most plants profitable utilization of gas and reliable operation was obtained. In contrast to these positive experiences the engines at Braunschweig's gas utilization plant were seriously damaged by corrosion, Dernbach (1985). The reason for this corrosion was the relatively high content of chlorinated hydrocarbons in the landfill gas. Total contents of 600 - 700 mg Cl/m^3, measured with a TOX-Analysator, were found in the gas.

In cooperation with Bergbau Forschung GmbH, Essen and sponsored by Federal Ministry for Research and Technology, Germany, investigations at the landfill site Braunschweig were carried out to evaluate the adsorption of chlorinated hydrocarbons from real landfill gas. The adsorption capacities and elimination rates from 9 activated carbons were studied.

2. Trace contamination of landfill gas

A large number of chloro- and fluoro-hydrocarbons (halocarbons) are included in the gas from several landfill sites in the Federal Republic of Germany. Some of these substances are toxic and responsible for corrosion in the engines. The reason for high contents of these components is the deposition of industrial refuse. Tab. 1 shows the range of values from halocarbons in several landfills in Germany, Rettenberger (1987).

Trichlorfluormethan	CCl_3F	1-84
Dichlordifluormethan	CCl_2F_2	4-119
Chlortrifluormethan	$CClF_3$	0-10
Dichlormethan	CH_2Cl_2	0-6
Trichlormethan	$CHCl_3$	0-2
Tetrachlormethan	CCl_4	0-0.6
1.1.1-Trichlorethan	$C_2H_3Cl_3$	0.5-4
Chlorethan	C_2H_3Cl	0-264
Dichlorethen	$C_2H_2Cl_2$	0-294
Trichlorethen	C_2HCl_3	0-182
Tetrachlorethen	C_2Cl_4	0.1-142
Chlorbenzol	C_6H_5Cl	0-0.2

Tab. 1: Chlorinated hydrocarbons in landfill gas in mg/m^3, based on airless landfill gas, Rettenberger (1987)

The mean value of the total chlorine content is 90-120 mg Cl/m^3, but values up to 300-400 mg Cl/m^3 are expected. At some landfill sites values up to 1.000 mg Cl/m^3 were found, Rettenberger (1987).

3. Experimental procedure

The investigations were performed in an adsorption plant on the landfill site Braunschweig shown in Fig. 1.

The landfill gas streamed through the cooler and the H_2S-adsorption to the halocarbon adsorption reactor. In this reactor the halocarbons were adsorbed by activated carbon. The total contents of chlorine in the inlet and outlet stream were semi-continuously measured using a modified AOX-Analysator. After reaching a content of 50 mg Cl/m^3 in the outlet gas the adsorption process was interrupted. The activated carbon was regenerated with hot steam (130°C) for 1 h. After a drying phase with nitrogen the next adsorption cycle followed. To reach steady state 7 to 10 adsorption/desorption cycles with each adsorbent were carried out.

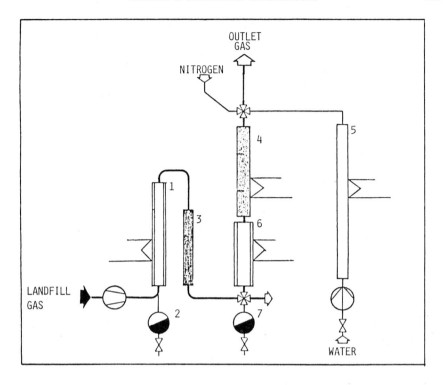

Fig. 1: Schematic view on the adsorption plant
 1: Cooler   5: Steamgenerator
 2: Condensate   6: Cooler
 3: H_2S-adsorption reactor   7: Desorbate
 4: Halocarbons adsorption
    reactor

The conditions of adsorption and desorption are presented in Tab. 2.

Adsorption	1. stage H_2S-elimination	2. stage Halocarbon-elimination
Length of the reactor (mm)	1000	1500
Diameter of the reactor (mm)	50	50
Temperature (°C)	16-20	16-20
Stream velocity (m/s)	0,3	0,3
Through put (m³/h)	2,1	2,1
Contact time (s)	3,4	5,1
Volume (l)	2	3

Desorption

Temperature	(°C)	-	130
Quantity of steam	(kg/h)	-	0,72
Stream velocity	(m/s)	-	0,2
Desorptiontime	(h)	-	2

Tab. 2: Conditions of adsorption and desorption

The composition of the landfill gas Braunschweig can be seen from Tab. 3.

Halocarbon	mg/m^3
Dichlordifluoromethane (F12)	5,7
1,1,1-Trichloroethane	2,8
Trichloroethylene	35,7
Tretrachloroethylene	115,2
Vinylchloride	6,2
1-1-Dichloroethane	2,3
cis-Dichlorethylene	379,0
trans-Dichlorethylene	8,1
Dichloromethane	2,7

Tab. 3: Composition from the landfill gas

The total chlorine content decreased in the last 2 years from 660 mg Cl/m^3 to 470 mg Cl/m^3. In this period the investigations to examine the adsorption capacities and elimination rates from 9 activates carbons were leaded through.

4. Results of experience

With every adsorbent it was possible to reduce the total chlorine content in the outlet gas below 20 mg Cl/m^3 by elimination rates greater than 95 %. With some activated carbons outlet concentrations below 10 mg Cl/m^3 were available. Considerable differences were established by the adsorption capacities. The results are shown in Tab. 4.

Activated carbon	Elimination rate (%)	Adsorption capacity (percentage by weight)
D 47/3 and C 40/3	93-97	1,06
D 55/2 and C 40/3	97-99	0,51
D 43/3 and C 40/3	96	1,09
Braunkohlenfeinkoks	96	0,20
SC XII 6x14	96-99	0,72
F 30/470	96	0,45
ED/47	96	0,59
Supersorbon K	98	1,28
Sorbonorit 3	96-99	1,13

Tab. 4: Elimination rates and adsorption capacities from the adsorbents

The results of a test series with activated carbon "Supersorbon K" from Lurgi GmbH are exemplary shown in Fig. 2. The inlet and outlet concentrations were semi-continuously measured during the adsorption time.

The course of the breakthrough curves were similar by the different activated carbons. In every test series no excessive variation in the outlet concentration during the adsorption time were established. The breakthrough happened very steep. After one hour the inlet concentration was reached.

In every case the operating time decreased after the first adsorption/desorption cycle approximately at 50 %, but remains almost constant in the cycles 3 to 10.

5. Conclusion

On the landfill site Braunschweig field trials were carried out to evaluate the adsorption of halogenated hydrocarbons from real landfill gas. The adsorption capacities and elimination rates from 9 activated carbons were studied. Using an adsorption process it was possible to reduce the total chlorine content from approximately 550 mg Cl/m^3 below 20 mg Cl/m^3.

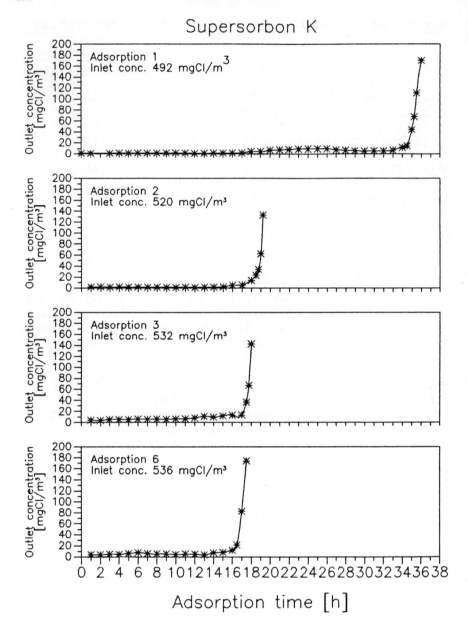

Fig. 2: Breakthrough curves from one activated carbon by different adsorption cycles

No excessive variation in the outlet concentration during the adsorption periods were established.

The adsorption capacities varied in a range from 0,2 to 1,28 percentage by weight. These values are strictly dependent on the composition of the landfill gas.

Literature

Franzius, V. (1987) Stand der Deponiegasnutzung in der BRD, Beiheft zu Müll und Abfall, Heft 26, 1987

Dernbach, H. (1985) Landfill gas utilization in Braunschweig - quality of gas and damages due to corrosion, Waste Management & Research (1985) 3, 149 - 159.

Rettenberger, G. (1987) Trace composition of landfill gas, International Sanitary Landfill Symposium, Cagliari 1987

Laboratory scale test for anaerobic degradation of municipal solid waste

Chr. Wolffson

1. Introduction

In Germany about 70% of occurring municipal solid waste (MSW) are deposited on landfills. The quality and quantity of landfill leachate are relatively well known. But it is difficult to measure gas production, so that the rates observed in different studies vary considerably.

In this research project we observed the influences of different MSW-components on the anaerobic degradation with the help of laboratory scale test cells. The knowledge about these connections is important for prediction quality of emissions, when different components are separately collected in order to recycle and therefore not deposited on a sanitary landfill. The operation technics as well as the technical installations have to be adapted on changed quality and quantity of leachate and gaseous emissions.

2. Experimental Procedure

The laboratory scale test has been developed to simulate the anaerobic processes including all running phases, the "Acid Phase" as well as the "Methanogenic Phase". The reactors are standard MSW-containers with 120 l volume (Fig. 1). These containers were covered with gastight PVC plates with integrated supplies for collection of all gas produced. Leachate was collected in small tanks (2.5 l) below the containers. 8 times a day the leachate was pumped back to the top of containers through a distribution system. Twice a week 1.5 l of liquid were changed with freshwater. The concentration of organic and anorganic components of gas and leachate was analysed for achieving decomposition phase and load of emission. To start

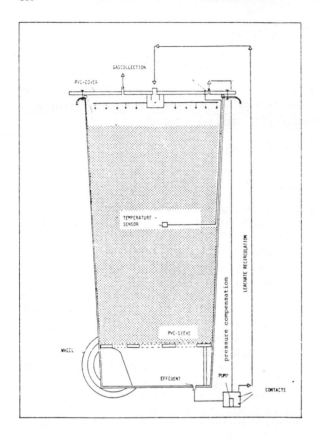

Fig.1: Cross section of test reactors

a test, the containers were filled with a mixture of 1 part crude MSW and 2 parts composted MSW and brought to a water content of 65%. This procedure is important for enhancing the biological processes. To the base material different MSW-fractions like newspaper, food waste, garden waste and sundry residues from waste recycle plants were added to. Containers were placed in a thermostatic room in that temperature of 30°C was maintained during the whole experimental time.

3. Results of experiments

3.1 Development of anaerobic degradation

The progressive curves of anaerobic MSW degradation has been achieved by means of approximately 100 tests in series with different objects. Typical re-

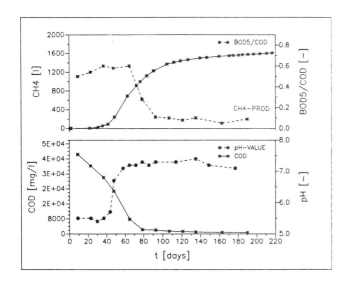

Fig.2: Typical leachate content and methane production relative to time

sults of gas production and leachate quality are presented in Fig. 2. When pH-value begins to increase the initial phase with high organic pollution without gas production has finished. In following we observed the highest gas production at that time when the increasing pH-value shows the turning point. When the highest organic pollution in leachate is occurring the BOD_5/COD-ratio is beginning to decrease. When this ratio amounts to ≤0,1 the so called "Methanogenic Phase" is starting. But at that time more than 60 % of whole gas-production has been observed in nearly all test series. This is an important point for an effective gas utilization is aspired to. It is finished when BOD_5/COD ratio has decreased. So very intensive enhancement techniques could result in a too short "Acid Phase" with following weak gas production. The stabilisation of waste, that means the reduction of organic Substrate, could be limited.

Solids analyses

Solid analysis were carried out before and after runtime of biodegradation test. Data of fresh MSW differ in a wide range between each series.
Nevertheless there was a strong relationship between volatiles on the one side and COD and carbon on the other side [1]

COD (g/kg) = 78.4 + 1.04 x VS (g/kg)
C   (g/kg) = 89.8 + 0.60 x VS (g/kg)

These correlations were the same with fresh MSW and stabilized MSW after the end of each experiment. Fig. 3 shows specific rates of biodegradation for different MSW-components. During runtime of test reactors only 35% of organic substrate of MSW has been reduced, although all processes of biological

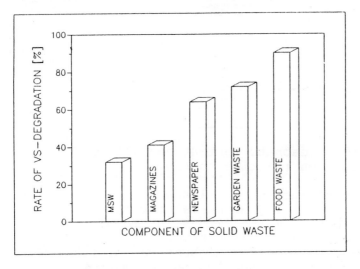

Fig.3: Specific rates of biodegradation for different MSW-components

degradation, gas production as well as changing leachate quality, has been finished. That means, that after the time of active biological activities in a sanitary landfill a considerable potential of organic material remained. Looking to specific components the rate of decomposition increases in succession of biodegradability: magazines, newspapers, garden wastes and mixed food wastes.

Leachate Quality

With the help of analyzing leachate quality the influences of different MSW components could be described. Fig. 4 contains the progressive curve of BOD_5/COD-ratio for test reactors in one series.

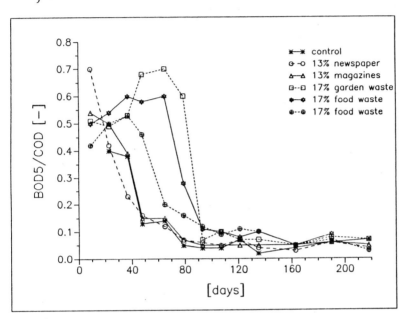

Fig.4: BOD5/COD-ratio for testcells with different additions to base material

A control reactor was filled with base material. To this different components were added. The figure shows, that the period of "Acid Phase" is extended in container 5 and 6 (food waste added) and 4 (grass added). But addition of paperfraction (container 2 and 3) effected in "Stabilized Methanoganic Phase" running only 40 days. Corresponding to that is the relation of the different carbon streams leaving the reactors. From control reactor 50 to 60 % of carbon are in leachate (TOC, alkalinity and CO_2) and 40 to 50 % are in gas (CH_4 and CO_2). Adding paper decreases the carbon stream in leachate and increases the carbon in gas.
The general goal of waste management is reducing mass of landfilling MSW while separating components like paper or organic material for composting. So

some testseries were carried out with residues fraction from waste recycle plants (without base material in testcells). We achieved that sorted waste containing only unimportant well biodegradable organic waste like paper, garden wastes, or vegetables run through same anaerobic phases of biodegradation like unsorted refuse too. Fig. 5 shows the development of gas-quality, but all emissions were clearly lower but measurable. A consequence for landfill technique could be that landfills filled with sorted MSW as well as those

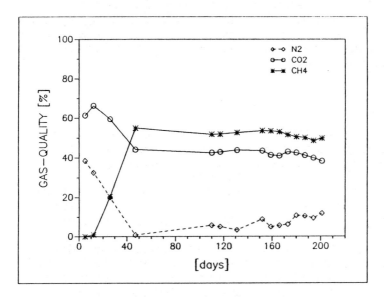

Fig.5: Gas quality of sorted MSW that contains only little biodegradable components

with unsorted MSW have their typical emissions. In any case orderly Gas collection and leachate treatment have to carry on.

Conclusion

Results of the test series show that the container test makes it possible to run through the processes taking place within landfills in a fairly short time of about 6 months. With the help of addition of different MSW-components we observed the gaseous and liquid emissions occurring at anaerobic degrada-

tion. In all cases about 60 to 70 % of whole
methane production was produced when BOD_5/COD-
ration decreased below 0,1. This would be a sign
for "Methanogenic Phase" in sanitary landfills.

The test series we carried out show a stabilizing
effect of paper fraction in MSW on the anaerobic
biodegradation. Recycling of this fraction could
result in a longer "Acid Phase" with high organic
pollution in leachate and weak methane production.
Simultaneous separation of well biodegradable fraction like garden waste or foodwaste is able to
equalize this trend. But also landfilling of sorted
MSW without well biodegradable components results
in gaseous and liquid emissions that require an
orderly technic for elimination.

Literature

1.) Ehrig, H.-J.,(1984), Laboratory scale test for
anaerobic degradation of municipal solid waste
Proc. of intern. solid waste and public cleansing
association Congress, Philadelphia
2.) Ehrig, H.-J.,(1980) , Beitrag zum quantitativen
und qualitativen Wasserhaushalt von Mülldeponien,
Veröff. des Inst. für Stadtbauwesen, T.U.Braun-
schweig, H. 26
3.) Wolffson, C.,(1985), Untersuchungen über den
Einfluß der Hausmüllzusammensetzung auf die
Sickerwasser- und Gasemissionen, Sickerwasser aus
Hausmülldeponien,- Einflüsse und Behandlung -,
Veröff. des Inst. für Stadtbauwesen,
T.U.Braunschweig,H. 39
4.) Mennerich, A., Wolffson, C.,(1986), Über-
legungen zur langfristigen Entwicklung und
Kontrolle von Emissionen bei Hausmülldeponien.
Fortschritte der Deponietechnik. Erich Schmidt
Verlag, Berlin

Session 5

Sludge Disposal and Landuse

LABORATORY EXPERIENCES ON THE FORMATION OF PERCOLATE FROM CIVIL AND INDUSTRIAL CONDITIONING SLUDGES

Marco BALDI*, Giorgio CHIERICO*, Vincenzo RIGANTI**, Michela SPECCHIARELLO**

* Water Research Center - University of Pavia
** General Chemistry Department, Chair of Commercial Chemistry - University of Pavia

Introduction

In Italy only a small percentage of solid wastes is treated by means of incineration; landfills accept more than 80% of urban and industrial wastes.

The methods for landfills construction and management are reported in the Deliberation of the Interministerial Committee dated July 27, 1984, in execution of the Decree of the President of the Republic, September 10, 1982, no. 915.

In landfills suitable for accepting both urban and industrial wastes the permeability, the retention and the absorption capacity of the clay layers between the mass of the wastes and the underground waters must be so that it could preserve these waters from pollution. The systems for draining and collecting the percolate, and as well as the ones for its treatment, must be such as to return to the environment surface waters complying with the quality limits set by Table A of the law dated May 10, 1976, no. 319 to the environment.

It is obviously important to predict, by means of suitable mathematical models, the qualitative and quantitative characteristics of the percolate vs the age of the landfill. The quality and the quantity of the produced percolate are mainly functions of the type and the rate of the decomposition processes which occur to the organic matter and to the water balance, which plays an important role on the contact time between water and solid wastes.

Many models have been described in the literature and the evolution of the characteristics of the percolate of many landfills has been analyzed[1] in the course of time, but the adjustment and the calibration of the proposed models still require further experimental research.

A first rudimentary previsional essay is constituted by the various elution tests provided for by law. In Italy two release tests, with water and acetic acid and with carbon dioxide saturated water, have been adopted, the latter being used for wastes characterized by inorganic composition. These tests have been subjected to severe criticism[2], since even some natural materials, free from contamination, have been proved to be unable to pass them.

Our previous works have shown that the test with water and acetic acid tends to overestimate the heavy metals concentrations in the percolate in short and mid-term, while the test with carbon dioxide saturated water tends to underestimate it.

More refined tests use column[4] or batch[5] elution processes; the comparison between tests performed on column and batch tests seems to point out that the obtained results do not differ substantially[6], except in the case of Cr(VI) release.

Anyway it is current opinion that the behaviour of an urban waste landfill (and more generally of a landfill mainly containing organic matter) may be better described in terms of batch reactor.

The anaerobic fermentation process which occurs in a landfill can be distinguished, as it is known, in an acid phase (in which the pH is lower than 7) and in a methanogenic phase in which the pH rises above neutrality.

The pH of a landfill poor in organic substance can instead remain for a long time at values above neutrality[7].

During the acid phase there is an increase in the solubilization of heavy metals, in particular of iron and manganese; however, their rate of concentration decreases, in the course of time. The concentration of other metals is generally lower, with irregular fluctuation: as, for instance, in nickel, chrome and copper.

Experimental Section

The aim of this laboratory research is the study of the percolate heavy metals concentrations vs imbibing time, which constitutes the necessary previous statement to a more extended programme directed to the adjustment of previsional models of landfill percolate quality and quantity.

It employs, for this purpose, a batch methodology, similar to the one proposed by D. C. Wilson[5] to investigate the behaviour of two types of sludges: anaerobic sludge produced by a municipal waste-water treatment plant and a mixture of sludges from various industrial processes characterized by high concentration of nickel, chrome, iron and manganese. Their analysis is reported in Table 1.

Table 1 - Composition of examined sludges

Parameter		Bio-conditioning sludge	Industrial sludge
Residual 105 °C	%	44.8	89.3
Ashes 600 °C	%	26.0	57.3
Fe	g/kg d.m.	9.95	114.4
Al	g/kg d.m.	17.9	0.9
Mn	g/kg d.m.	0.23	14.7
Cr	g/kg d.m.	0.21	33.5
Ni	g/kg d.m.	0.12	12.4
Cu	g/kg d.m.	0.25	0.29

Six samples, (200 g) of both sludges, were passed through a sieve and introduced in sealed teflon containers containing 200 cm^3 of imbibing media constituted by the following solutions:
- carbon dioxide saturated water, to simulate the situation in which the methanogenic fermentation process takes place;
- 0.05 molar solution of acetic acid buffered at pH 5 with caustic soda, to simulate the situation in which the synthesis of short-chain fatty acid occurs in the landfill;
- 3.75 g/dm^3 EDTA solution, to simulate the presence of strong complexing agents. Each container was stirred for ten minutes to ensure a good contact between liquid and solid phase surface; afterwards one series of containers was stored at 21 °C and another identical series was stored at 60 °C. At preset times a part of the content of each container was sampled, in order to avoid the alteration of the solid/liquid ratio; the sampled material was filtered through a 0.45 μm membrane and the filtrate, after acidification with HNO_3 Suprapur, was analyzed with a Spectraspan IV argon plasma continous-current spectrophotometer Spectrometric Inc., Mass. By means of this instrument it is possible to perform multielementary analysis on the same sample with good precision and accuracy and with a minimal matrix effect.

Results and discussion

The laboratory research was carried out for 37 weeks; it was impossible to draw liquid after the 30th day from the bio-conditioning sludge samples kept warm, and the same occurred after 20 weeks to industrial sludge samples kept warm.
The pH of the liquid phase, initially acid around pH 5 in the samples treated with water-acetic acid and CO_2 saturated water and substantially neutral in the samples treated with EDTA solutions, becomes rapidly alkaline (pH 7.5-8.0) in the industrial sludge samples stabilized by means of $Ca(OH)_2$. In the bio-conditioning sludge samples the pH becomes slightly alkaline after a few days of contact and remains so until the end of the experiment.

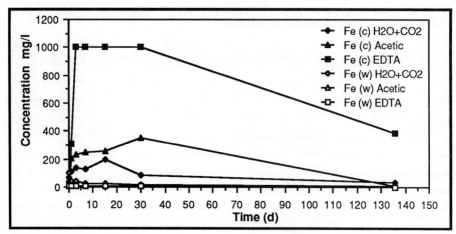

Figure 1: Fe concentration vs contact time into elutes coming from industrial sludge.

Iron is contained in large amounts in the industrial sludge and, with a smaller but significant concentration, in the bio-conditioning sludge. It is easily extracted from cold industrial sludge sample by EDTA solution; in the course of time the concentration tends to decrease for all the eluents. Smaller amounts of it are extracted both from the cold bio-conditioning sludge and from the heated one except in the presence of EDTA.

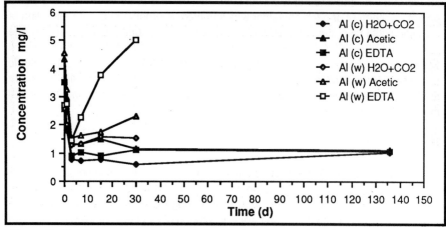

Figure 2: Aluminum concentration vs time into elutes produced by municipal sludge

Bio-conditioning sludge contains much higher aluminum amounts than the industrial one and this fact seems to play an important role on the long term elutes concentrations, which do not significantly depend on the temperature and on the nature of the imbibing media, but mainly on its relative amounts in the sludges. As it should be possible to observe in Fig. 2, heated bio-conditioning sludge sample had released higher aluminum amounts than the cold ones mostly for EDTA solutions.

Large manganese amounts are available in the industrial sludge while bio-conditioning sludge contains this metal at low concentration. The behaviour of manganese is similar to iron: it is easily extracted from the industrial sludge warm samples by carbon dioxide saturated water, but in the course of time concentration in the solution tends to decrease. Smaller amounts are extracted from the bio-conditioning sludge: the best solvent is EDTA, both for cold and heated solution. As for industrial sludge, long-term concentrations are lower than the short-term ones.

Chromium concentration in industrial sludge is two magnitude orders above its concentration in bioconditioning municipal sludge; but, contrary to aluminum, the elutes metal amounts do not seem to be function of that. As it is shown in Fig. 3 the most important phenomena is the chromium release from the cold industrial sludge mixture in presence of EDTA solution. Chrome concentration in EDTA elutes, coming from warm industrial sludge sample and from both bioconditioning sludge, are low and similar to the values obtained by means of other imbibing media; this is probably due to lower complex stability or to competition phenomena.

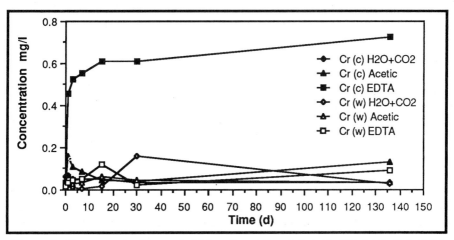

Figure 3: Cr concentration vs imbibing time into elutes coming from industrial sludge.

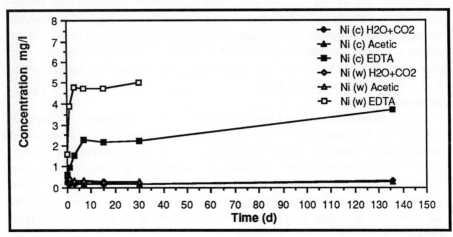

Figure 4: Nickel concentration vs imbibing time into elutes coming from municipal sludge.

Like chromium, also nickel concentration, in bio-conditioning sludge is lower than in the industrial one. Nickel is released by cold industrial sludge at high concentration in presence of EDTA solution while carbon dioxide saturated water is a good solvent at 60 °C; its behaviour is similar to chromium, but its complex stability constants are higher than the chromium ones. This is pointed out by Fig. 4 where we can see that significant heavy metal amounts are also extracted from the bio-conditioning sludge by EDTA containg solution in despite of sludge low content.
The other solutions do not extract significant amount of nickel from both sludges, either cold or warm.

Copper concentrations in elutes and percolates are described as mostly variable, this is probably due to its capacuty to build up complexes with various complexing agents. In the two tested sludges its concentration is of the same order of magnitude. The maximum release values were found in the bio-conditioning sludge after few days of imbibing time, but at medium and long terms all concentrations were very close.

Conclusions

Iron, aluminum and manganese show the peculiarity characteristics of yelding almost everytime a concentration peak at short term, the values of which are generally much higher than the long-term one. When cold, after 260 days the concentration of the examined metals in every contact solution except the ones containing a

strong complexing agent such as EDTA, stabilizes in a relatively narrow range of values. The solution containg EDTA which simulates the presence of strong complexing agents, extracts much higher concentrations with respect to the other solutions in the following cases:
- from the industrial sludge, chrome and nickel when cold;
- from the bio-conditioning sludge, iron, manganese and nickel when cold and iron and manganese when heated.

A good solvent for manganese and iron in industrial sludge is also costitueted by the solution containing carbon dioxide.

The results as a whole:

i. confirm the ability of the carbon dioxide saturated solution to extract from the sludges significant amounts of manganese and iron;

ii. point out the importance of the role of complexing agents on the heavy metals release;

iii. indicate that in the absence of strong complexing agents even considerable diffrences in the composition of the sludges do not lead to great differences in metals concentrations into the elutes at long term.

iiii. show that a higher temperature facilitates the retention of imbibing media mostly by the bio-conditioning sludge.

References

(1) Erhig, J. H. (1983). Waste Management and Research, 1, 53-68
(2) Canepa, P., Mazzuccottelli, A. (1985). La Chimica e l' Industria, 67 (1-2), 3-9
(3) Specchiarello, M., Riganti, V., Baldi, M. In Acque Reflue e Fanghi (Ed. A. Frigerio)Vol. I, 551-558. Centro Scientifico Internazionale, Milano.
(4) Genon, G., Marchese. F. (1987). Abstract of Sixth International Conference "Chemistry for Protection of the Environment" , Torino, 201.
(5) Wilson, D.C.et al. (1982). Environmental Science and Technology, 16 (9), 847-853.
(6) Jackson, D.R. et al. (1984). Environmental Science and Technology, 18 (9), 668-676.
(7) Brunner, P.H., Moench, H., McDow, S. (1983). E.A.W.A.G. News, 22/23, 17-19.

TREATMENT OF SLUDGE IN AN LGW PLANT, SYSTEM VØLUND *)

Mogens Fritze

INTRODUCTION

In 1931 Vølund delivered the world's first continuous waste incinerator plant in Gentofte near Copenhagen. This was actually the beginning of a continuous development, a working area which has gradually changed its character from being a system for destruction of waste into an advanced waste-to-energy system.

Until a few years ago focus has only been set on the solid waste, i.e. household and industrial waste. Other waste products in a liquid state were of less interest, because this waste could be stored within the area of the factory by virtue of a substantial smaller volume. This took place more or less by legal ways of burying of the waste, or the waste was transported to refuse dumps which at that time were <u>real</u> dumps, i.e. without any kind of control.

BASIS

In 1969 VØLUND MILJØTEKNIK carried through a series of tests with sludge incineration at Gentofte incinerator plant. The results were published in a form of a very extensive documentation. Unfortunately, the activities of VØLUND and other companies in this field did not result in building of plants. At the beginning of 1980, VØLUND MILJØTEKNIK started working again with the problem. VØLUND MILJØTEKNIK had a long-winded target of total treatment of the many kinds of waste from reception to flue gas cleaning. Some of the specific areas embraced treatment of manure and sludge which was started after some pre-examinations in a VØLUND mini plant with a capacity of approx. 25 kg per hour. In 1981 these examinations resulted in building of a pilot plant with a capacity of approx. 300 kg sludge per hour.

Sludge is being drained in many ways by traditionally known methods, and there is a close connection between increasing water content and the parallel connected costs of construction and working.

VØLUND MILJØTEKNIK has examined this circumstance and found the most suitable field for the LGW system. Figure 1 shows traditional drainage methods with expected dry matter (DM) and H_2O content. The hatched area indicates the field of application of the LGW system.

*) We have decided to apply for patent as well as to register the design as LGW (Low Grade Waste) - System VØLUND - with the specific characteristic that the system is suitable for treament of waste with a high water content.

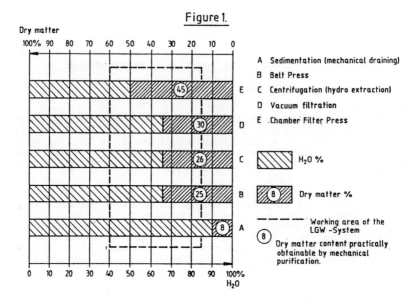

Figure 1.

CHOICE OF SYSTEM

The purpose of the LGW system is reception of sludge after final drainage. At an early stage the components for the entire process are being selected. Besides, it is estimated whether recirculation shall be used during the process, and whether pre-examinations or tests will be necessary.

An LGW system can be used in several ways:

1. Sludge is conveyed directly to a furnace for waste incineration (see figure no. 4).
2. Sludge is conveyed indirectly to a furnace for waste incineration (see figure no. 3).
3. Drying of sludge and incineration can take place in an LGW system placed on a purifying plant, perhaps with power or combined power and heating production.
4. The LGW system can be used as a pure drying plant for production of a saleable final product.

Recirculation is being used if the sludge cannot enter the press directly. Therefore, dried sludge is being added to the raw sludge in order to obtain a suitable content of dry matter.

To illustrate the recirculation quantity during various operations figure 2 shows an ideal state of the plant in equilibrium with curve A, while curve B and curve C, respectively, show start-up and stopping.

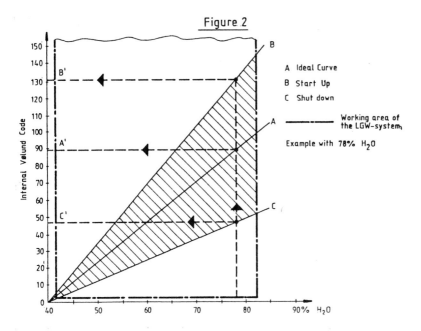

Figure 2

A Ideal Curve
B Start Up
C Shut down

—————— Working area of the LGW-system

Example with 78% H_2O

The character of the sludge cannot be determined from the content of dry matter alone. It depends on the fiber content, the quantity of unorganic and organic materials, possible grease content and share of salts which are especially found in the chemical and pharmaceutical industry. Besides, lime is being used in many of these industries and to a certain extent in the municipal sewage disposal plants, too, which can be traced in the sludge.

FUNCTION (system with indirect drying) figure 3

The sludge is being received in the incineration plant in a reception hall with gates. These gates shall always be closed during emptying of the containers. The silo is equipped with a turnable lid. It is equipped with sucking out meaning that possible inconveniences are being minimized by the right way of operation. The sludge is being pumped with a mono or a piston pump depending on the content of dry matter, to a pre-storage silo placed over the press unit.

The purpose of having a pre-storage silo is the wish for an exact dosing of wet sludge to the press unit as well as to make it possible to inspect the equipment in front during operation. The pre-storage silo doses the sludge out into a screw conveyor which at the same time acts as a mixing conveyor for wet and recirculated, dry sludge. The dosing of the sludge takes place automatically via the control which is being determined by the exhaust temperature from the cooling outlet of the post-drying unit.

The press unit is placed vertically in a pre-drying unit which is a lined chamber, especially framed. The screw conveyor leads the sludge into the press unit where it is being compressed, structuralized, and pre-dried at a high speed. Half of the water content vaporizes, and the flue gas temperature decreases from approx. 1050° C to approx. 550° C. The pre-dried sludge is conveyed via a special flap-system (*) into a combined post-drying unit with a cooler which, in principle, is built-up as a drying drum with stationary inlet and outlet. Especially during start-up and less during operation, there will be a small amount passing the flaps (1-3% of the amount introduced) which will be discarded during start-up, i.e. it will be led back to the sludge silo. In an equilibrium state of the plant it will be carried to the outlet for dried material, if no demand for a sterile product has been made.

(*) We have applied for patent and registered the design.

Figure 3 shows a typical LGW plant in connection with a VØLUND incineration plant. This will be the most common arrangement in connection with new plants.

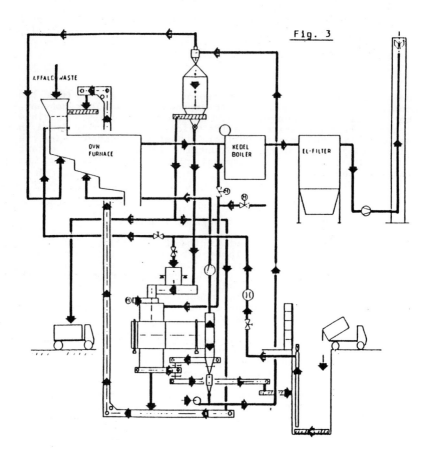

Fig. 3

Dried material which runs out of the cooler of the post-drying unit with a temperature of 50-100° C is carried by a transport system to a storage silo for dried material with a volume adjusted to the method of operation of the plant. 3 levels are built into the silo. When the upper level is reached, the silo will be emptied to the middle level of dried material. This is then being transported to the feeding chute in an adapted speed. During normal operation the storage silo cannot be emptied below the middle level, as it is necessary to have dried, recirculated material for start-up.

Dried material is taken out for recirculation from the area between the middle and upper level.

The entire LGW system is encapsulated and vacuumed in order not to contribute to the environmental problems by taking in a new type of waste into a traditional incineration plant.

Figure 4 shows an LGW system with direct feeding into the furnace top. It will be advantageous to use this arrangement by building it in onto existing waste incineration plants, or at places where the sludge/waste proportions are reasonable, i.e. approx. 1:4.

However, the system does not have the same advantages as an indirect system. It is characterized by being relatively easy to establish, and the plant investment is reasonable.

Many components in the direct and indirect system are the same, thus giving one the possibility to change a direct system into an indirect later on, provided that the space is present.

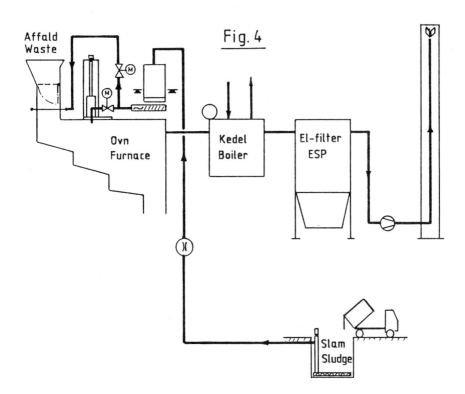

WASTE CONDITIONS

As an illustration of the connection between number of press units and sludge figure 5 shows the curves which are normative by the temporary choice of method of working, size of plant, size of modulus and the draft of the EDP programme needed.

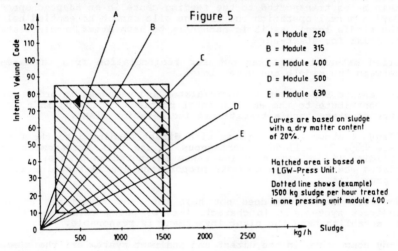

Figure 5

A = Module 250
B = Module 315
C = Module 400
D = Module 500
E = Module 630

Curves are based on sludge with a dry matter content of 20%.

Hatched area is based on 1 LGW-Press Unit.

Dotted line shows (example) 1500 kg sludge per hour treated in one pressing unit module 400.

ENERGY CONDITIONS

To illustrate the relations between the calorific value of the waste, the content of dry material, H_2O and the ratio sludge/waste and their influence on one another, figure 6 shows how an incineration plant can be charged considering the state of the sludge. Furthermore, our calculations provide for a variable operation. In periods with surplus of heat, the surplus can be used for drying of sludge. The sludge is then taken to a storage silo for use in periods of peak load. The operation can be varied during all 24 hours, for instance a 8/16 time adapted to the circumstances in week-ends.

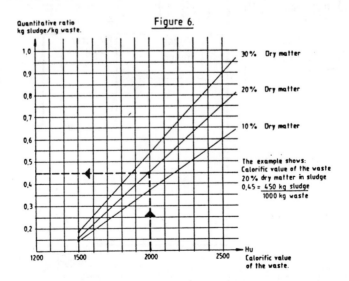

Figure 6.

The example shows:
Calorific value of the waste
20% dry matter in sludge
$0,45 = \frac{450 \text{ kg sludge}}{1000 \text{ kg waste}}$

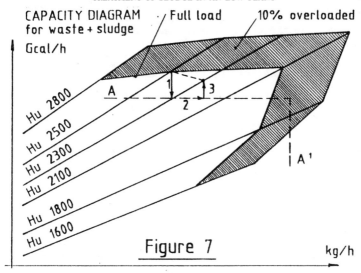

Figure 7 (capacity diagram)

From curve A it can be read how the drying and incineration of sludge are affected directly. The total upper calorific value (example Hu 2500) falls to curve A, but the total amount of combustible waste is increased to a certain extent because the dry matter of the sludge after drying contributes positively, both on the mass and energy side of curve A¹.

The diagram can be explained in this way:

1. Evaporation of water results in a decreasing average calorific value.

2. The total amount of dry matter is increased by adding sludge to the waste.

3. Increased calorific value because dried dry matter in the sludge will again increase the average.

The LGW system can also be used as a pure drying system. This means that by adding micronutrients to a waste product, i.e. sludge, the dried product will have some characteristics meaning that it can be used directly by the end user. We have carried out an extensive test as mentioned previously, and can apply the results to other products.

EXPERIENCE

Finally, we will study the results gained in practice during extensive tests with various products of waste:

1. Test with manure from farming, sludge, etc. (1982/83).

2. Sludge from 3 sewage disposal plants in the neighbourhood of Copenhagen, with or without lime (1983).

3. Waste from tanneries. Preliminary test in 1984 leading to an extensive test in January/February 1987.

4. Sludge from 2 sewage disposal plants in the neighbourhood of Copenhagen, with or without lime. The test was paid by the customer and was concluded by test in the field and pot experiments (1985).

5. Sludge from a Danish chemical factory. The test was paid by the customer and concluded by a test in the field (1985).

6. Waste from slaughterhouses with or without sludge. The test was paid by the customer (1986).

7. Waste from tanneries. Extension of the test mentioned in section 3. The test was partly financed by trade organization, the Ministry of Environment, advisers, and VØLUND MILJØTEKNIK (1987).

8. Marketing, pilot studies, contract (1986-1987).

9. First order for a plant in France with a capacity of approx. 700 kg sludge/h. Working time 7600 hours/year. (1987).

10. Start-up of the first LGW plant - system VØLUND. Spring 1989.

11. Test at the incineration plant I/S REFA (Denmark) with a direct system max. 900 kg sludge/h.(2nd half of 1987).

12. Building of 2 full scale plants at I/S REFA and I/S FASAN for direct and indirect drying and incineration of sludge (1988-1989).

ECONOMY

Revenues

If the present charges for storing of sludge in dumps are used, a plant with for instance 1000 kg sludge per hour/7600 hours per year may have the following revenues:

Charge to the dump
    7600 tons - DKK 47.00/t   =   DKK 357,200.-   ($ 60,000.-)

Waste charge
    7600 tons - DKK 40.00/t   =   DKK 304,000.-   ($ 50,000.-)

Revenues - excl. VAT   =   DKK 661,200.-   ($ 110,000.-)

provided that the charges for transport are the same as for the dump.

If a cooling system is used, one can expect a reduced amount of coolers because the dried sludge can be stored in silos and be used for incineration in peak periods.

Revenues are not taken into consideration as the dried materials are being regarded as a zero contribution to the incineration process, i.e. you "borrow" flue gas from the incineration furnace and return it in a lower level together with an amount of dried material.

The structure of figure 4 is a standard LGW plant with special consideration to the water content of the sludge which, in this case, is on the verge of the possibility of the system. It is built together with a new incineration plant, and the price of construction amounts to approx. DKK 8 mill, exclusive of buildings. Continuous operation is anticipated, i.e. 7600 hours per year, having the necessary staff on the day shift, i.e. 1 man for operation and maintenance (316 days of 8 hours a day = 2533 hours per year).

ADVANTAGES OF THE LGW SYSTEM

- Small energy consumption during the process compared with other pill-producing processes which press a dried product.

- Low energy consumption for the evaporation of water because of a big surface when structurizing the sludge in thin threads.

- Storage of energy by means of pills which can be transported without caking and with a little formation of dust.

- The equipment can be used both for indirect and direct drying, thus giving a greater flexibility.

- Great amount of sludge in proportion to waste (1:4) for the direct method and (1:2.5) for the indirect method when having 20% dry matter as a basis.

- Great span in water content in the sludge (90% - 50%).

- The system can be built onto existing plant.

- The direct system does not require must space.

CONCLUSION

Today, VØLUND MILJØTEKNIK A/S is in possession of a very modern draft for treatment of sludge and other wet waste with a big working area compared with other systems for drying of sludge and incineration.

VØLUND MILJØTEKNIK is of the opinion of having the right product at the right moment seen in the light of the measures taken all over the world to clean the waste water in a better way. This, naturally, will result in large amounts of sludge which can be treated in an LGW plant - System VØLUND.

The LGW system is constructed with a view to improvement of existing fluidized-bed-systems, furnaces etc. and is suitable for forming part of other plants for treatment of

WASTE-TO-ENERGY

REFERENCES

In connection with the performed tests following reports have been published:

1. Drying of Animal Manure with a View to Combustion
   (H. Have, Royal Veterinary and Agricultural University, and M. Fritze, Vølund Miljøteknik A/S).
   1980.

2. Drying of Chicken Manure with a View to Combustion
   (H. Have, Royal Veterinary and Agricultural University, and M. Fritze, Vølund Miljøteknik A/S).
   1981

3. Heat Extraction from Animal Manure by Combined Drying, Combustion, and Heat Recovery
   (H. Have, Royal Veterinary and Agricultural University). 1984.

4. Conversion of Wet Biomass, Peat, and Waste by Combined Drying and Combustion
   (H. Have, Royal Veterinary and Agricultural University, and V.H. Pedersen and M. Fritze, Vølund Miljøteknik A/S). 1984.

PRODUCTION OF AN EARTHLIKE MATERIAL BY COMBINING SEWAGE SLUDGE WITH POROUS MINERAL SUBSTANCES

D. HOLTZ, Umweltamt, Darmstadt, W. Germany

The sewage sludges which are continuously created by waste water purification are predominantly used for agriculture after appropriate stabilization in liquid form or brought to dumps or burned after treatment or partial dehydration with chamber filter presses, screen belt presses or centrifuges, respectively, with subsequent conditioning with burnt lime. The use of sewage sludges for agriculture is only possible outside of the vegetation period, that is maximally eight months in the year. The amount of sludges that can be applied per hectare and year is limited by the Sewage Sludge Rules in effect in the Federal Republic of Germany since June 25, 1982 (Bundesgesetzblatt 1982, Part I, p. 734 and following), among others, so that a total supply of agriculture and the return of valuable fertilizer and humus-forming substances to the natural cycle is practically impossible, even if the other boundary conditions of the Sewage Sludge Rules are observed. The methods for treatment of sewage sludges in preparation for dumping or burning require high technological expenditure with considerable investment and operating costs with additional using up dump capacity under loss of economically relevant amounts of fertilizer and humus materials.

The problem was to find a method for the production of a storable and applicable soil improvement agent using aerobically or anaerobically stabilized sewage sludge which enables year-round sewage sludge disposal. In contrary to dumping or incineration of sewage sludges, a technically and economically usable soil improvement agent should be obtained, so that the fertilizer and humus materials in the sewage sludges could be reclaimed for soil improvement.

At present, the only existing production facility is in Darmstadt, West Germany. It was at the Darmstadt facility that all research and development, toward producing the substrate (trade name ORGABO), took place. The Darmstadt plant uses anaerobically digested sewage sludge, which is partially dewatered with the help of a centrifuge. The resulting paste-like material is then mixed with oil shale slag. This slag is run through a screening device which separates all material too large for the production. Testing has shown that lava, pumice and similar materials are also suitable for the production of ORGABO. A major advantage to the ORGABO concept is that the production facility can be built for continuous operation, and using technical equipment which has long since proven dependable in practice (i.e. centrifuges or screen belt presses, conveyor belts, double roll mixer etc.). With the exception of a flocculation agent used in the sludge dewatering process, no further physical or chemical treatment is necessary for the production process. The end result is a soil improvement agent or cultivation substrate, which retains the humus and fertilizing elements of the sludge in their natural form. Percolation tests have shown that the sludge particles bound to the mineral carrier substance are stable against wash-out effects, due to the application of flocculation agents during the dewatering process. After production the material is ready for use. Contrary to compost materials no storage period for additional rotting is necessary, due to the fact that the decomposition of the organic shares in the sludge is biologically completed during sludge digestion or aerobical stabilization, respectively.

ORGABO as an end product has much the same characteristics as a good top soil. After a very short storage period, it is either odorless, or has a weak earth-like smell. The substrate has the following properties:

- pH value   approx. 6.0
- Humus content   3 - 5   %
- Water storage capacity   35   % / wt
                                 57   % / vol
- Total nitrogen   0.24   %
- Water soluble nitrogen   0.08   %
- Phosphorus   1.00   %
- Potassium   0.10   %
- Calcium   2.30   %
- Magnesium   0.50   %

According to the composition of the sludge used, the fertilizing elements content can vary minutely. In respect to heavy metals content, only sludges which meet the sewage sludge regulations of the Federal Republic of Germany are used.

The major part of the water soluble fertilizing elements are eliminated by the sludge dewatering process. **The remaining fertilizing agents are more** or less high molecular compounds which are decomposed by bacterial activity over a long period of time. Thus, the biological decomposition products become available for plantlife and additional fertilization is not necessary for years.

In addition to the property of storing the fertilizing elements until needed by the plants, the substrate is, as a technical product, free of stones, loam and weeds, and allows the necessary air and water to penetrate. Furthermore, due to its structure, the ORGABO material is stable against compaction even under great mechanical loads, such as heavy farming equipment. The material has also been tested on slopes, roads and embankments, and has proven stable against erosion even during heavy rainfall.

Moreover ORGABO has been extensively tested by the agricultural research and testing institute of the state of Hessen, F.R.G., and their research departments in the cities of Kassel-Harleshausen and Darmstadt. These tests were conducted over a three year period (1983 - 85), to establish the following:

- Fertilizing effects
- Percentages of increased yield
- The effects of heavy metals content, and their possible transition to the plantlife

In accordance with these test results, yield increases were achieved as follows:

- oats   19 % increased yield
- oat straw   7 % increased yield
- rape seed   34 % increased yield

The tests were conducted with diverse substrata including sand, and variable percentages of ORGABO. These results were measured in the second test year (1984), with no further soil treatment or fertilization before the vegetation cycle. The content of toxic heavy metals (lead, cadmium and mercury), in **weidel-grass (Lolium perenne), was measured in the** first test year (1983), and found to be exceptionally low.

These test results and several years experience with large scale field trials, indicate a distinct, slow release fertilizing effect, due to microbial release of nitrogen, phosphorus and other compounds. In bacteriological and hygienic testing, no evidence of pathogenic germs could be found. A bacteriological comparison was made between ORGABO and normal arable soil. The study showed that ORGABO was perfectly suitable as a substratum for the soil's microflora. As a result of its findings, the agricultural research and testing institute of the state of Hessen, F.R.G., gave its approval for ORGABO to be used in all forms of landscaping and recultivation.

Since November 1984 the ORGABO substrate is available on the West German market. In the meantime, more than 4o.ooo tons of the ORGABO material have been produced and used in recultivation of landfills, of roadsides and in landscaping. Presently tests are being conducted with tree saplings and shrubbery, at particularly difficult locations, in down-town Darmstadt. The recultivation projects showed a stable and thick growth rate in a very short time. ORGABO has also been proven to effectively stop erosion. The original test site (1983) of approximately 2 hectares is still showing a strong, healthy growth, and has not been fertilized or treated in any way since the test began, five years ago.

The city of Darmstadt was granted a patent for the manufacturing process (DE 3409274 C 1 from 28-2-85). International patents are pending.

Publications

1. Patent: DE 3409274 C 1 from 28-o2-1985
2. Holtz, D.: (1984), Bodenverbesserer aus Faulschlamm und Ölschieferschlacke, WLB - Wasser, Luft und Betrieb 4/84, 48 - 5o
3. Holtz, D.: (1985), Bodenverbesserer, Wirtschaftsraum Darmstadt 85, 65 - 68, published by: Magistrat der Stadt Darmstadt
4. Holtz, D., Fellner von Feldegg, S.: (1987), Klärschlammverwertung mit dem ORGABO-Verfahren, in Recycling von Klärschlamm 1 (Ed. K.J. Thomé - Kozmiensky,U. Loll), 46 - 52, EF-Verlag für Energie- und Umwelttechnik, Berlin, ISBN 3-924511-18-7

LIMING OF SLUDGE FROM BIOLOGICAL TREATMENT OF COKERY WASTEWATER

E.S. Kempa and T. Marcinkowski

1. Introduction

Agricultural uses of sewage sludges have become the most common method of their disposal. Sludges generated by coking plants include large amounts of nutrients. But these cannot be recovered or recycled until the cyanides and phenols present in the sludge are removed or chemically fixed, and the pathogens occurring there are completely (or, at least, partly) destroyed. In an earlier study (Kempa et al. 1985) a disposal method was developed for sludges from a cokery-owned sewage treatment plant. The method made use of slaked lime (suspension in aqueous solution or hydrated lime) and quicklime.

2. Experimental methods and procedures

The experiments were run with sludge samples (referred to as Sludge A, Sludge B and Sludge C), which had been dewatered to a dry solids content of about 13 wt.%. Liming by the quicklime method involved 6.5 dm^3 volume vacuum flasks. Temperature was recorded continually in a compensatory system coupled with resistance thermometers. Treatment procedures using lime suspensions or hydrated lime were carried out in 8 dm^3 volume tanks. Physicochemical, bacteriological and parasitological analyses were conducted prior to, and following completion of, the liming procedure. Proportions of the sludge components were varied.

3. Results

3.1. Physicochemical, bacteriological, and parasitological composition of raw sludges

Physicochemical analyses were to determining the presence of fertilizing substances (N,P,K,Ca; MLVSS,) and toxic species (CN^-, phenols). The results are in Table 1.

Table 1

Physicochemical Composition of Raw Sludges

Substance or parameter	Unit	Sludge A	Sludge B
Water content	%	96.12	95.34
MLVSS	% dry sol.	81.76	70.23
Total nitrogen	% dry sol.	12.30	7.33
Phosphates	% dry sol.	0.81	1.10
Potassium	% dry sol.	0.54	0.50
Calcium	% dry sol.	1.50	1.57
Cyanides	% dry sol.	0.027	0.030
Phenols	% dry sol.	0.001	0.002
Ether extract	% dry sol.	0.54	0.51
Water extract	pH	6.71	6.40

Bacteriological and parasitological analyses included the following determinations: counts of mesophilic bacteria and resting spores, as well as the number of parasitic worm ova, protozoa and arthropoda. The data of interest are in Table 2.

Table 2

Bacteriological and parasitological composition of raw sludges (1 cm^3 volume samples)

Pathogen	Sludge A	Sludge B
Resting spores	$1.0*10^4$	10^1
Mesophilic bacteria	$9.0*10^8$	$9.8*10^6$
Worms (ova)	5	5
Protozoa (cysts)	absent	absent
Arthropoda (ova)	<300	10 to 12

3.2. Liming

Tables 3 and 4 give physicochemical, bacteriological and parasitological composition of the sludges after lime treatment.

A number of samples were treated with 5 % CaO, 10% CaO, and 20 % CaO in aqueous solutions (the doses applied amounting to 5 g CaO/dm^3 sludge, 10 g CaO/dm^3 sludge and 20 g CaO/dm^3 sludge, respectively). Each test was repeated three times.

Sludge treatment with pulverized hydrated lime involved two doses of calcium hydroxide: 50 g and 150 g CaO/dm^3 sludge. The test was repeated three times for each dose.

The treatment process involving pulverized quicklime was conducted with seven doses ranging from 150 to 1500 g CaO/dm^3 sludge. It is a well-established fact that, owing to the moisture contained in the sludge, calcium oxide liming is an exotermic process.

Table 3

Chemical composition of sludges after liming

No. of sample	Sludge	Lime dose g CaO/kg dry sol.	Cyanides	Phenols	Ether extr.
		Lime suspension in water *			
1	A	39.6 (5%susp.)	0.007	0.0008	0.58
2	A	79.1 (10%susp.)	0.005	0.006	0.38
3	A	118.7(20%susp.)	0.004	0.005	0.36
4	B	41.1 (5%susp.)	0.012	0.0010	0.22
5	B	82.1 (10%susp.)	0.009	0.0010	0.14
6	B	123.3(20%susp.)	0.008	0.004	0.12
		Pulverized hydrated lime *			
12	A	410	0.004	0.0002	0.30
13	A	1220	0.003	0.0001	0.11
14	B	400	0.007	0.006	0.10
15	B	1230	0.004	0.0003	0.08
		Pulverized quicklime **			
21	A	1500(temp.338K)	0.002	0.0002	0.16
22	A	150(temp.306K)	0.005	0.0006	0.32
23	B	1500(temp.340K)	0.002	0.0003	
24	B	500(temp.309K)	0.005	0.0010	0.32

* Determined after 6 weeks of storage.
** Determined immediately following completion of the process.

It was anticipated that the liming process would yield the destruction of pathogenic organisms, as well as a partial removal of cyanides and the fixation of phenols with calcium hydroxides.

Table 4

Bacteriological and parasitological composition of sludges after liming (1 cm^3 vol. samples)*

No. of sample	Mesophilic bact.	Resting spores	Worm ova	Arthropoda ova
Lime suspension in water				
1	$1.2*10^5$	$3.0*10^2$	single	absent
3	$9.0*10^2$	10^1	absent	absent
4	$1.7*10^4$	10^2	single	single
6	$1.1*10^3$	10^1	absent	absent
Pulverized hydrated lime				
12	10^2	20	single	absent
13	0	5	absent	absent
14	20	30	absent	single
15	0	0	absent	absent
Pulverized quicklime				
21	0	5	absent	absent
22	$3.1*10^6$	40	single	absent
23	$1.4*10^1$	1	absent	absent
24	$1.0*10^6$	$2.0*10^4$	absent	absent

* Lime doses as in Table 3.

4. Summarizing comments

Lime treatment was effective, specifically when the CaO doses made the sludge temperature rise to about 340 K. The product displayed sufficiently low concentrations of phenols and cyanides, as well as a sufficiently good pasteurization effect. Analyses of water extract from the sludge-lime mixtures show that phenols persisting in the sludge are insoluble, because they are fixed with Ca^{2+} ions to phenolates. Thus cokery sludges treated via this method may be used for reclamation of sour and degraded soils.

5. Conclusions

1. Sludges treated with calcium oxide suspensions in aqueous solutions or hydrated lime show a satisfactory degree of dewatering, thus enabling a troubleless storage in lagoons for many years.

2. Quicklime treatment is preferable as it enables the following: rapid removal of some part of the cyanides present in the sludge, chemical fixation of phenols, and biological inactivation. Containing considerable amounts of calcium, nitrogen, phosphorus and potassium, the product is fit for agricultural uses. It applies to sour soils, but may also be used for the reclamation of degraded land.

References

Kempa, E. Marcinkowski, T. Szpadt, R. (1985). Multi-variant disposal concept for sludges produced by a cokery-owned biological sewage treatment plant. Report of the Institute of Environment Protection Engineering, Technical University of Wrocław. PWr I-15/SPR-42/85. (Unpublished, in Polish).

APPLICATION OF SLUDGE-LIME MIXTURES TO SOIL RECLAMATION : VEGETATION TESTS

T. Marcinkowski

1. Introduction

1.1. Effect of calcium on some properties of soil and vegetation.

It is a well-known fact that calcium ions are washed down to deeper soil levels, thus accounting for the impoverishment of the top soil. The presence of calcium improves the air and water conditions in the soil, reduces acidity, stimulates humidifying processes, increases nutrient availability, and exerts a favourable effect on the degradation of organic matter. The contribution of calcium to vegetation growth is considerable (e.g., to the formation of cell walls). Any discontinuation in calcium supply may inhibit crop growth. Since calcium absorption is irreversible, continuous supply of this species in adequate quantities so as to ensure appropriate plant growth has become a necessity.

2. Objective and scope of the study

One of the research groups at the Institute of Environment Protection Engineering, Technical University of Wrocław, has concentrated for many years on the decontamination of sewage sludges by

lime treatment to make them applicable as fertilizers. Concurrent vegetation tests (pot tests) have been carried out in five series on several classes of soil (Marcinkowski 1983) to establish appropriate sludge-lime mixtures (digested sludge).
The study involved soil samples of three productivity classes (Class I, Class II and Class V) and three pH levels (7.2, 5.2 and 4.2, respectively) with oat (*Avena sativa* L.) as a testing plant. Fertilization efficiency was established for each class by measuring the leaf length and determining the green and dry weights of the plants.

3. Method of pot testing

The pots had a diameter and a depth of 133 mm and contained 1 dm^3 volume of soil. Moisture content (80% of maximum absorbability) and ambient temperature (between 296 and 298 K) were measured throughout the tests. The pots were illuminated with 800 lx light for 14 hours a day. The time span for vegetation was 21 days. Soil portions were mixed with appropriate sludge-lime doses. Sludge application was repeated three times for each sample. Sludge was dried at 378 K and disintegrated in a porcelain mortar before it was mixed with the respective soil portion.
Each pot was planted with 50 oat seeds and sprayed with distilled water. To reduce evaporation, the top soil was covered with a thin sand layer. Aeration was from the bottom through glass pipes. Each testing series included a control with no sludge-lime treatment.

4. Sludge-lime doses

Sludge-lime dose calculations involved the calcium ratio criterion, using the measured values of exchange and hydrolytic acidities (Nowosielski, 1980). Table 1 gives sludge doses determined for individual productivity classes. These quantities refer to 100% of the CaO component and have been determined for the weight percent of CaO in the mixture. In every instance, the sludge dose applied was an integer multiple of the base dose (Dp; 2Dp; 3Dp, etc.).

Table 1. Base doses for the soil tested

Acidity removed	CaO dose (kg/ha) for natural soil		
	Class I	Class II	Class V
hydrolytic	1088	2330	3224
exchange	34	866	1203

5. Results

Series I was run on natural soil of Class I with pH=7.2. The CaO/sludge (dry solids) ratio in the fertilizing mixture was 5/2. The mixture doses ranged from 1,500 to 6,000 kg/ha. Leaf length increased with the increasing dose up to the value of 4,500 kg/ha. From there, leaf growth was less pronounced - at the 6,000 kg/ha dose the growth effect was almost identical to that at the 1,500 kg/ha. There was also an increase in pH with the increasing dose - from 7.2 in the control to 8.1 when the maximum dose had been applied. These

variations fall in the tolerable pH range for oat.
Series II involved Class II soil with pH=5.2, and a ratio of components in the mixture as in series I (the mixture dose varying from 3,500 to 1,4000 kg/ha). The best results were achieved with the 10,500 kg/ha dose. The maximum dose yielded a pH=7.6. Analyses of soil and crops have shown that the quantities of calcium and magnesium available to the plants were sufficient, but those of NPK were to poor.
In the tests of series III the NPK deficiency was overcome by additional fertilization with the following doses : 576 kg N/ha; 72 kg P/ha and 576 kg K/ha. The best effects were achieved with the base fertilizer dose of 7,000 kg/ha. Thus, the application of additional doses including NPK reduced the optimum mixture dose by a half. The soil sample treated with the maximum fertilizing mixture dose reached a pH of 7.9.
The tests of series IV were run on sandy soil of Class V productivity with pH=4.2. The CaO/sludge ratio in the mixture was 5/2, at a dose varying from 5,000 to 30,000 kg/ha. Vegetative mass increment was the highest at the 10,000 kg/ha dose. When the maximum dose was applied, the increment in the vegetative mass amounted to only 55% of that in the control. This inhibition of crop growth (which had been noted at the 15,000 kg/ha dose along with the flavescence of leaves) should be attributed to the overliming of the soil. Thus, pH increased from 4.2 in the control to 8.5 at the maximum fertilizing mixture dose. It is worth remembering that at pH>=8.0 the availability of

Table 2. Vegetation tests involving Oat (*Avena sativa* L.) on natural soils fertilized with CaO/dry solids=5/2 mixture
(* = optimum dose)

Series productivity	Dose of fertilizer	pH after testing	Relative value compared to control (%)		
			leaf length	green mass	dry mass
1	0	7.2	100	100	100
1	1500	7.6	112.6	128.4	112.8
1	3000	8.0	191.9	161.2	146.0
1	4500*	8.1	141.5	166.3	175.2
1	6000	8.1	128.2	130.1	109.9
2	0	5.2	100	100	100
2	3500	6.5	122.3	131.4	133.9
2	7000	7.1	130.0	174.8	158.4
2	10500*	7.5	140.9	199.7	180.0
2	14000	7.6	128.9	150.1	151.1
3	0	5.2	100	100	100
3	3500	6.1	117.9	118.8	113.3
3	7000*	6.6	141.1	125.9	116.2
3	10500	7.0	131.6	112.8	109.7
3	21000	7.9	89.5	66.7	82.2
4	0	4.2	100	100	100
4	5000	6.0	115.8	106.7	110.8
4	10000*	6.9	125.3	119.0	115.9
4	15000	7.6	110.5	100.1	105.6
4	30000	8.5	73.7	58.7	52.3

nutrients becomes limited. The results of series I through series IV are given in Table 2.

Series V involved the same soil samples as series IV, but the CaO-sludge ratios were 3/2 and 1/1.

Table 3. Vegetation tests involving oat (*Avena sativa* L.) on natural soils fertilized with a CaO/dry solids = 3/2 and CaO/dry solids = 1/1 mixtures

Soil productivity	Dose of fertiliz. kg CaO /ha	CaO/ dry sol.	pH after testing	Average leaf length cm	Average green mass g/dm^2	Average dry mass g/dm^2
Class I	4000	3/2	7.8	22.3	11.84	0.513
Class II	8000	3/2	7.3	14.7	7.88	0.483
Class V	8000	3/2	7.0	12.8	5.02	0.329
Class I	4000	1/1	7.6	22.9	12.63	0.571
Class II	8000	1/1	7.2	16.3	8.35	0.576
Class V	8000	1/1	7.0	13.9	5.81	0.404

Samples with maximum increment in the vegetative mass are listed in Table 3. The best effects were obtained with the fertilizing mixture wherein the dry solids content had its maximum value.

5. Summarizing comments

Digested sludges treated with lime for decontamination exert either a stimulating or an inhibiting effect on plant growth. There exists a possibility to determine an appropriate fertilizing method for every soil-sludge-lime complex so as to improve the productivity class of the soil involved. In the study reported here the optimum range of the fertilizing mixture dose was very low (minimum value) for Class I soil with pH approaching 7.2. The optimum dose reached the maximum value for Class V soil (sand) with an acidic pH.

As far as the optimum dose is concerned, the following observations have been made. For Class I and Class II soils, the best effects are achieved with a fertilizing mixture dose which is three times the base dose (Series I and II). For sandy soil (Class V), the mixture dose should be twice as high as the base dose to yield optimum growth. The calculated base doses are sufficient for pH adjustment, but they are insufficient to cover the demand for nutrients. Hence, optimum growth conditions are not achieved until a significantly increased lime-sludge fertilizer dose is applied. It is the neutral pH that provides adequate availability of nutrients. This finding has been substantiated by the results of series III (with additional NPK fertilisation) and those of series V (the fertilizing mixture has been enriched with dry solids (sludge) in relation to CaO).

In each of the five series there was an overliming of the soil (at doses exceeding their optimal values) which contributed to the flavescence of the leaves and to the inhibition of growth.

References

Marcinkowski, T. (1983). Report of the Institute of Environment Protection Engineering, Technical University of Wrocław. Series SPR No. 63/83.

Nowosielski, O. (1980). Metody oznaczania potrzeb nawożenia. PWRiL, Warszawa.

Heavy Metals in Sewage Sludge: Concentration Variability and Elution Possibilities.

M.Ottaviani*, L.Olori*, G.Basile**
*Istituto Superiore di Sanità, Rome, Italy
**University of Naples, Italy

Introduction

One of the principal limitations for the use of sewage sludge in agriculture is the heavy metal content found in it. The risk for the cutering of such metals within the food chain either through groundwater contamination or by a greater assimilability from plants, depends on the concentration and the physical chemical form that they have inside the soil system. For these reasons, this work shows the results of a 0,5M acetic acid dynamic elution on a sludge coming from a municipal purging plant. The investigated metals were: Cd, Cu, Ni, Pb, Zn, Cr. Under a toxicological profile Cd represents the source of greater preoccupation, therefore, in addition to the total content it is essential to know, also, available Cd fraction for the plants. Among the factors that condition such disposability, clayed materials and organic substances have a pre-eminent role, this work refers, also, to the Cd absortion on imogolite (aluminium silicate polymer occuring within the mineralogical composition of vulcanic soil) and on purified humic acid.

Materials and Methods

The heavy metals presents on the sludge samples coming from a civil-wastes plant (Table 1) got into a solution with a binary acid mixture (HNO_3 and $HClO_4$) (1) and their content was determined by means of AA spectrophotometry with and without flame (Perkin Elmermod 460). The dynamic elution tests were carried out with a 0,5 M solution of acetic acid (2). After stratification, the heavy metals found in the liquid phase were determined as above. The obtained results are shown in table 2.

The Cd absortion on imogolite was estimated on imogolite obtained by synthesis according to the method described by Farmer (3).The purity of the product was controlled with the use of an x-ray diffractometer, electronic microscopy and I.R.spectroscopy.
The Cd absorption on imogolite was defined by shaking for 24 hours and successively centrifugating 20 mg of imogolite,with a 20 ml of a 0.05 N $Ca(NO_3)_2$ solution which contained Cd on a quantity able to realize concentrations between 0,05 and 0,06 µmoles/Lt.The pH's of the suspensions were taken to predetermined values maintained between 6,5 and 7,6.
The Cd concentrations contained in the clear centrifugated solution were determined by AA spectrophotometry without flame. The absorbed quantities from imogolite at the different pH values was calculated as the difference between the initial concentration and the one determined on the solution at equilibrium.
The obtained results are shown in Figure 1 were the quantities of Cd adsorbed by the imogolite (µmoles/Lt) are expressed,according to the diverse pH values,as a function of the Cd concentration within the solution at equilibrium.At the pH's values of 6,4-6,8-7,2-7,6 and for a concentration at equilibrium of 0,3 µmoles/Lt the quantity of Cd adsorbed from imogolite was equal to 0,08-0, 15-0, 25-0, 40 µmoles /g respectively.The Cd adsorption on humic acids was obtained by using humic acid extracted from brown soil of the vesuvian area.The humic acids were obtained by shaking for 24 hours a soil sample containing a 0,5 N NaOH solution (air-dried and decalcificated with a 0,1 H_2SO_4 solution),using a 1/10 soil solution ratio. After centrifugation and removal of the residual part, the supernatant liquor was acidified with a 1 N H_2SO_4 solution up to a pH 2 and left to rest for one night.The precipitated humic acids, separated by centrifugation,were re-dissolved in a 0,1 N NaOH solution and therefore,re-precipitated with a 1 N H_2SO_4 solution.After washing with a solution of HCl and HF (0,5% by vol.)in order to reduce the ashes content, the humic acids were centrifugated and dialyzated

against distilled water. Lastly the humic acids were
lyo philizated.
Cd adsorption by means of humic acids was defined by shaking for 24 hours and successively centrifugating 20 mg of humic acids, with a 10 ml 0,05 N solution of $Ca(NO_3)_2$ wich contained Cd on a quantity able to realize concentrations between 0,20 and 0,60 µmoles/Lt.
The pH of the suspensions were taken to predetermined values maintained between 3 and 7.
The Cd concentrations contained in the centrifugated solutions were determined by AA spectrophotometry without flame.
The obtained results are reported and figure 2,where the quantities of Cd adsorbed by the humic acids (µmoles/Lt) are expressed,according to the diverse pH values, as a function of the Cd concentration within the solution at equilibrium.
At the pH values of 3,0-4,0-5,0-6,0 and for a concentration at equilibrium of 0,2 µmoles/Lt the quantity of Cd adsorbed by the humic acids was egual to 0,03-0,07-0,13-0,26 µmoles/g respectively.

Conclusion

The dynamic elution tests in an acid environment for acetic acid (even if it is considered an extreme condition for elution,from an enviromental point of view, that could involve sludges when in contact with the environment) showed that a certain transfer occurs and has to be considered in those cases in which it is intended to perform the spreading on acid soil.
Futhermore,from the adsorption tests was found that the Cd adsorption by means of humic acids by means of imogolite is sensible to the pH;as a matter of fact, when testing with humic acids,the lowering of each pH unit in the considered interval,causes a 50% reduction of the adsorption while,when testing for imogolite, lowering the pH of about a unit determines A 90% reduction of the adsorption.

Acknowledgments

The Authors thanks Doctor Alfonso Iadaresta,Mr. Paolo Morgia and Mr Fabio Savelli for theirs cooperation.

Bibliography

1 - IRSA (1985) Analytical Methods for sludges, physical-chemical parameters.CNR,vol.64.
2 - Farmer, V.C.,Fraser,A.R.,Tait,J.M. (1977). Synthesis of imogolite:a tabular aluminium silicate polymer,J.Chem. Soc.Chem.Comm. 462-463.

TABLE N° 1

Plant characteristics and quality digested sludge

pH	7.4	
C_t	32.4	ab. eq. 12000
N_t	5.0	treatment type:water sludge activated
P_t	1.2	treatment type:sludge anaerobic digested
K_t	0.5	
Ashes	41.9	
SS	59.0	

TABLE N° 2

Heavy metals concentrations in sludge and in elution test with acetic acid

	Minimum value	Maximum value	Mean value	Eluate	% Eluate
Cd	0.5	1.5	1.1	0.08	7.2
Cu	240	450	310	15.5	5.0
Ni	60	90	72	14.0	19.4
Pb	120	212	170	8.5	5.0
Zn	650	1050	780	195	25
Cr	60	40	65	0.5	1

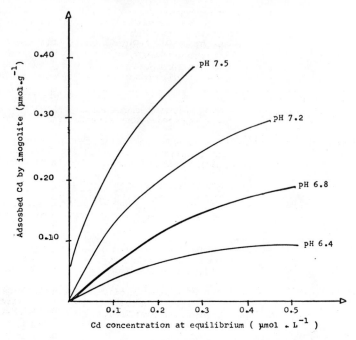

FIG. 1 - Isotherms for Cd adsorption by imogolite at 20 °C, at different pH.

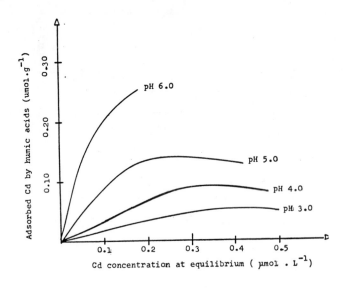

FIG. 2 - Isotherms for Cd adsorption by humic acids at 20° C, at different pH.

LONG-TERM FIELD TRIALS WITH SEWAGE SLUDGE IN COMPARISON WITH FARMYARD MANURE AND COMMERCIAL FERTILIZER

Kalju Valdmaa,
Swedish University of Agricultural Sciences,
S-750 07 Uppsala, Sweden

This investigation into the use of wastewater sludge on agricultural land started during the autumn of 1960 when the first sludge was spread. The first harvest year of this trial was 1961 and the series of trials is still continuing.

The investigation concerns the biological value of sludge in comparison with manure and commercial fertilizer under field conditions in normal agriculture. The experimental plan is as follows:

   A. No sludge, manure or commercial fertilizer
   B. N and K with commercial fertilizer
   C. P and K with commercial fertilizer
   D. N, P and K with commercial fertilizer
      1. No sludge, manure or fertilizer
      2. Half dose sludge = 4 and 8* tonnes DM/ha
      3. Half dose manure = 4 and 8* tonnes DM/ha
      4. Full dose sludge = 8 and 16* tonnes DM/ha
      5. Full dose manure = 8 and 16* tonnes DM/ha
      * = 8 and 16 tonnes application in Period V (autumn 1981).

According to the original experimental plan, sludge and manure were to be applied at the various doses to each fifth crop. For practical reasons, sludge and manure were applied 5 times (autumn 1960, 1965, 1971, 1977 and 1981) so that the respective periods in between applications were 5, 6, 5, 4 and 6 years. The crops grown in the trial corresponded to those used in the farm rotation. Consequently, the crops grown during each respective period vary widely.

During the total period of the investigation, 1961-87, a total of 26 harvests were taken, of which 20 were cereals and 6 were hay. Fertilizing with NPK commercial fertilizers was done annually at optimum doses for the respective crops.

In the autumn of 1981 the trial was redesigned by applying double doses of both sludge and manure so that the half doses corresponded to 8 tonnes of DM/ha and the full dose to 16 tonnes DM/ha. The wastewater sludge applied to the experimental field was obtained from the Uppsala Municipal Wastewater Works and the manure from the Uppsala Agricultural Society's experimental farm at Västerby. The sludge and the manure used in the experiment were originally only analysed for dry matter and concentrations of nitrogen, phosphorus and potassium. Complete analyses of both sludge and manure were started in Period III. The percentage of N in sludge and manure is fairly similar in all periods. The sludge varies between 3.1 and 4.4 % of dry matter and the manure between 2.4 and 2.7 % of dry matter. The phosphorus levels vary slightly in the sludge, between 2.1 and 4.4 % of dry matter, whereas in manure they vary between 0.5 and 0.8 % of dry matter. As regards heavy metals, the cadmium concentration had decreased from 4mg to 2mg per kg DM sludge at the last analysis.

The yields of the different crops in each experimental treatment have been converted into yield units/kg DM using the following factors. The yield of straw is not included in this presentation.

Concentration factors for conversion of crop yields into units/kg DM according to Helmenius et al. (1957).

Crop: Winter wheat   factor   1.17
     Barley   "   1.15
     Oats   "   1.00
     Hay   "   0.60

The yields obtained for the different years have been presented by periods, which implies that each manuring interval with sludge or manure corresponds to one period. These periods with recurrent applications could not be kept constant for practical reasons and varied from 4-6 crops in the different periods.

Effects of sludge, manure and commercial fertilizer

Figure 1 shows the yields during different periods for commercial fertilizer, sludge and manure.

The effect of the treatment with sludge and manure without commercial fertilizer was positive throughout for treatments 2 and 4 for sludge and for 3 and 5 for manure, in comparison with treatment 1. Relatively large yield-promoting effects were obtained during the second period in treatment 4, with a full dose of sludge (18 % ± 8.27 %) and for the same treatment in period 4 we find corresponding values of 18 % ± 2.5 %, which is statistically significant (Valdmaa, 1986). Corresponding significances were also obtained in treatments 2, 3 and 5 in periods IV and V.

Mean values for the 26 harvests show that sludge had a slightly higher effect when given in full dose than the corresponding effect with manure. However, the differences were frequently minimal and within the standard error.

When studying the effect of commercial fertilizer in treatment B (N+K), in treatment C (P+K) and in treatment D (N+P+K), without sludge and manure, it is observed that the effect in treatment C can be compared with the yield level given in treatment A, which is

Fig. 1: The effect of treatment with sewage sludge, manure and commercial fertilizer (NPK) in units of yield/ha in means during the five periods, 1961-1987.

Independent of commercial fertilizer

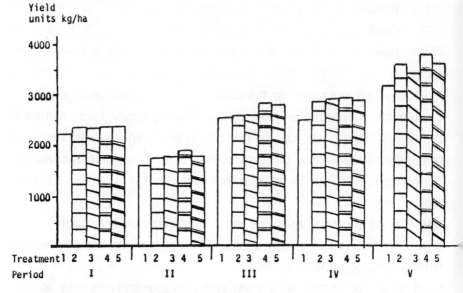

Independent of sewage sludge and manure

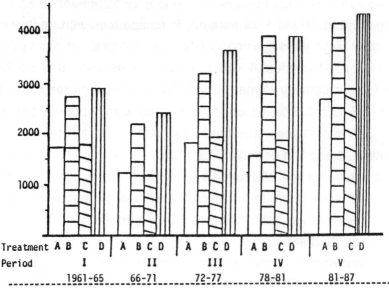

1 = no sludge or manure
2 = 4 and 8^x tonnes DM/ha sludge   4 = 8 and 16^x tonnes DM/ha sludge
3 = 4 " 8^x "   "   manure   5 = 8 " 16^x "   "   manure
x = period V

within the limits of the standard error in periods I, II and III. However, there is a significant effect in treatment C during the fourth and fifth periods, probably as a result of the sludge and manure not corresponding to the phosphorus requirement of the crops grown and possibly also to the potassium requirement. On average for the 26-year-period, the P+K effect in commercial fertilizer is no larger than the standard error of the experiment during the experimental period.

The addition of N+K in treatment B and N+P+K in treatment D gave significant yield-promoting effects. This, in turn, provides evidence of the good utilization of the commercial fertilizer by the plants and that the fertilizer supplied in the sludge or manure did not cover the nutrient requirements of the crops.

Analytical results

In order to examine the uptake of cadmium in plants, analyses of vegetation samples were made on the last crop, barley, in the fourth period as well as on the first, second, third and fourth crops in the fifth period, with doubled doses of sludge and manure.

Samples of vegetation were taken at the heading or panicle stages of the crops and samples of hay at harvest. The analyses of vegetation samples (Table 1) show that the cadmium uptake by the last crop in the fourth period in 1981 in treatment A was the same, being below the detection limit. In treatment B, with simultaneous application of commercial fertilizer N+K, the concentrations of cadmium were higher in all treatments. In treatment C, with P+K, the uptake of cadmium in all treatments was of similar size, but with slightly higher levels in treatment C1 (P+K alone) and in treatment C5 (manure and P+K). In treatment D, with N+P+K, the cadmium uptake in treatment D1 (N+P+K alone) and in treatment D5 (full dose of manure with N+P+K) was slightly lower than in treatments D2 and D3 and treatment D4 (full dose of sludge with N+P+K).

Table 1: Cadmium Cd µg/kg d.m. in vegetation[x] samples from a trial with sewage sludge and manure in comparison with fertilizer NPK.

Treatment	Barley 1981	Barley 1982	Oats 1983	Barley 1984	Hay 1985 I cut	Hay 1985 II cut
			µg Cd/kg DM			
A1 no fert., sludge or manure	<10	<10	17	13	27	16
A2 no fert., + 1/2 sludge	<10	15	17	10	-	-
A3 no fert., + 1/2 manure	<10	20	8	18	-	-
A4 no fert., + 1 sludge	<10	28	35	31	30	15
A5 no fert., + 1 manure	<10	29	19	20	28	15
B1 no sludge or manure + NK	45	11	43	25	-	-
B2 1/2 sludge + NK	28	19	25	27	-	-
B3 1/2 manure + NK	24	16	16	17	-	-
B4 1 sludge + NK	49	38	24	22	-	-
B5 1 manure + NK	46	<10	15	30	-	-
C1 no sludge or manure + PK	14	<10	14	18	-	-
C2 1/2 sludge + PK	11	26	19	13	-	-
C3 1/2 manure + PK	<10	<10	13	12	-	-
C4 1 sludge + PK	<10	37	26	17	-	-
C5 1 manure + PK	19	<10	17	13	-	-
D1 no sludge or manure + NPK	34	29	17	19	43	27
D2 1/2 sludge + NPK	43	31	21	28	-	-
D3 1/2 manure + NPK	42	19	15	21	-	-
D4 1 sludge + NPK	46	<10	28	37	51	40
D5 1 manure + NPK	29	13	14	23	41	25

x= Plant cut off at soil level in early heading stage.
-= Not Analysed
1981 barley is the fourth annual crop (4 years after treatment) following applications of sludge and manure: 1/2 sludge and manure = 4 tonnes DM/ha
1 sludge and manure = 8 tonnes DM/ha
1982 barley is the 1st year´s crop with applications of:
1/2 sludge and manure 8 tonnes DM/ha
1   sludge and manure 16 tonnes DM/ha

The cadmium uptake in the first and subsequent crops in the fifth period largely had the same pattern whereby in some cases with doubled doses of sludge and manure there was increased uptake of cadmium. For example, treatments A4 and A5 in 1982, 1983 and 1984 and in hay cut I and hay cut II the uptake is as high as in treatment A1 without sludge and manure.

In treatment B with N+K, the cadmium uptake was higher in treatments B2 and B4 and in C2 and C4 than in the other treatments during 1982. In the same year in treatment D with N+P+K, the double application of sludge and manure (D4 and D5) decreased the uptake of cadmium. In the second year after sludge and manure applications the cadmium uptake was higher in all treatments with commercial fertilizer (2 and 4) than in the manure treatments (3 and 5) apart from B1 with N+K without sludge and manure where the highest cadmium uptake occurred, which might have been a coincidence.

In the third year after sludge and manure applications (1984), the cadmium uptake was much the same in the different treatments apart from treatment D4 (full dose of sludge) which showed the highest level of cadmium (37 µg/kg DM) which might also be a coincidence.

In the fourth year after sludge and manure application, the analyses have only comprised treatments A1, A4, A5 and D1, D4 and D5 in both harvests. The analyses show that when commercial fertilizer (N+P+K) was applied, the cadmium uptake increased in all treatments with N+P+K in comparison with treatment A. The treatment with sludge (D4) also shows increased uptake.

SUMMARY

These long-term experiments comparing manure and commercial fertilizer (N+K, P+K and N+P+K) reveal that sludge increased the yield as much as manure. This effect of sludge was improved in some cases when supplies of nitrogen were withdrawn.

Higher yields were obtained, on average, throughout the entire experimental period (26 harvests) in treatments with full doses of sludge (totally 48 tonnes DM sludge/ha), the half doses of sludge (totally 24 tonnes DM sludge/ha) as well as in full doses of manure (totally 48 tonnes DM manure/ha), which gave slightly higher yield effects, whereas the lowest was given by the half dose of manure (totally 24 tonnes DM manure/ha).

Measurements of the uptake of cadmium in vegetation samples taken from the last crop of the fourth period as well as the crops in the fifth period show no significant differences between sludge and manure. However, there is a slight increase in cadmium uptake in treatments with sludge which was strengthened in some treatments following application of commercial fertilizer.

REFERENCES

Helmenius, A., Rydå, K. & Woldmar, G. 1957. Husdjursskötsel (Animal Husbandry)

Valdmaa, K. 1986. Long-Term Field Trials with Wastewater Sludge in Comparison with Farmyard Manure and Commercial Fertilizer. Recycling in Chemical Water and Wastewater Treatment. Schriftenreihe ISWW Karlsruhe Bd. 50, 271-282.

Session 6

Hazardous Waste

NEW ZEALAND'S HAZARDOUS WASTE MANAGEMENT STRATEGY

Margaret L Bailey, Ministry for the Environment,
Wellington, New Zealand

Introduction

The safe disposal of hazardous wastes created by technologically advanced societies is a challenging task. New Zealand has not had the serious problems faced by other countries. However, steps are being taken to improve New Zealand's waste management practices.

In June 1984 the Interdepartmental Working Party on the Management of Hazardous Wastes (IWP) was set up to report to government on the management of hazardous wastes. In May 1985 the Working Party's report (IWP, May 1985) containing a national strategy for managing wastes was released for public comment. A revised version of the Working Party's report was prepared the following year (IWP, April 1986).

The Secretary for the Environment assumed responsibility for the report when the Ministry for the Environment was set up in December 1986. It was rewritten after being referred to the Interagency Co-ordinating Committee on Hazardous Substances and subsequently released as "A Strategy for Managing Hazardous Wastes" (IWP, Aug 1987).

The Strategy

The strategy proposed that the prime responsibility for the management of hazardous wastes rested with regional and local government, with technical support and financial incentives provided by central government. The essential elements of the management strategy were the following four stages:

Stage 1: A survey of wastes, including hazardous wastes, and current management practices in each region. This includes examples of and opportunities for waste re-use and reduction, low-waste technology, and treatment to make wastes less hazardous.

Stage 2: An investigation of existing and potential landfill sites in each region to determine their suitability for hazardous waste treatment and disposal.

Stage 3: The preparation of a management plan in each region to state what wastes will go where and under what conditions, in consultation with industry and other interested parties.

Stage 4: The implementation of the management plan by agencies identified in the plan, in accordance with defined operational procedures.

Central government provided money to encourage regional authorities to undertake waste surveys and site investigations and to prepare waste management plans. To date, 19 out of 22 regions have started implementing this strategy. Eight surveys of wastes have been published. A subsidy scheme has been initiated for approved hazardous waste management works.

Problems in Waste Management

Unfortunately the strategy has not solved all the problems in managing wastes. Incidents continue to happen, as the following examples from late 1987 to early 1988 show:

* Drums containing toxic wastes were dumped at the Silverstream tip, a large landfill serving part of the Wellington region. The engineer in charge of the tip did not know what they contained. This followed another incident where a worker was affected by fumes when a drum was accidentally punctured.

* Birds have been found covered in oil at another tip in the Wellington region after swimming in a pool of wastes containing oil, septic wastes and contaminated water.

* A load of isocyanates was turned away from a tip run by the Auckland Regional Authority because of inadequate documentation. The waste did not return, and considerable effort was required to locate it. A local solution is being considered to prevent a similar occurrence.

* A transformer containing PCBs exploded at a sub-station, spreading PCBs around the immediate environment. The area was cleaned up, and the contaminated material is now awaiting export for disposal, as there is no suitable facility for disposal of this material in New Zealand.

In addition, many landfills continue to be poorly operated, particularly in relation to the disposal of hazardous wastes, and some are sited in inappropriate locations. Training of operators and supervisors is needed.

A New Policy

Hazardous wastes are a problem because costs and benefits of their production and disposal are not shared equitably between those who generate them and the community in general. There is no realistic incentive for waste generators to lower their waste output as they do not bear the real costs of disposal. These costs, mainly environmental degradation and health risks, are borne by the community.

A hazardous waste management strategy should aim to eliminate this inequity. If waste generators are made to meet the full costs to society of disposal, the marginal cost to them of producing waste increases. As incentives for businesses aim to balance marginal costs against marginal benefits, hazardous wastes should be produced at a level which minimises costs to the community. This is in keeping with the "Polluter Pays Principle" (OECD, 1975). Under a "polluter pays" regime, waste reduction and resource recovery techniques are likely to be adopted by waste generators.

Any strategy should be consistent with the objectives of economic efficiency, equity, good environmental management and practicality, as agreed to by the New Zealand Government in its review of the resource management statutes (those dealing with planning, mining, water and soil conservation, hazardous substances, and so on).

The structure of local and regional government is also under review, with likely impacts on waste management. The fundamental principle is that local or regional government should be selected for a particular task only when the net benefits of such an option exceed all other institutional arrangements. Although some government involvement is inevitable in the management of hazardous wastes, the allocation of responsibilities between the tiers of government in the current strategy will need to be reviewed once the outcome of the local government review is clear.

Consequently, at the end of March 1988, the New Zealand Government agreed that a policy should be prepared based on the following broad principles:

(a) Production of hazardous wastes should be reduced, avoided or eliminated by the use within practicable limits of waste reduction techniques.

(b) The use of practicable resource recovery and re-use techniques should be maximised.

(c) The responsibility for making satisfactory provision for hazardous wastes rests with the generator, in keeping with the Polluter Pays Principle, provided that efficiency, equity and practicability are not compromised.

(d) Due consideration should be taken of the need for public participation and of Maori cultural values.

The formulation of the new hazardous wastes management strategy is to be undertaken by officials from five government departments and co-ordinated with the reviews of the resource management statutes and of local and regional government. The Ministry for the Environment can undertake interim measures for dealing with specific wastes if urgent action is required. At present the only wastes in this category are PCBs, and work on a strategy for managing PCBs is in progress.

Further work

Other work is being undertaken by the Hazardous Wastes Task Group, a multi-sector group which reports to the Interagency Co-ordinating Committe. This group, convened by the Ministry for the Environment, has set itself a list of priority tasks in the management of wastes. Its first priority, the management of PCBs, has resulted in a draft report on a strategy for managing PCBs (PCBs Core Group, March 1988). It is now investigating what legislation is needed, and is undertaking research into incineration facilities and waste reduction and low waste technologies.

References

IWP (Interdepartmental Working Party on the Management of Hazardous Wastes) (May 1985). "Draft Report", Ministry of Works and Development, Wellington.

IWP (April 1986). "Final Report", Ministry of Works and Development, Wellington - unpublished.

IWP (Aug 1987). "A Strategy for Managing Hazardous Wastes", Ministry for the Environment, Wellington.

OECD (Organisation for Economic Co-operation and Development) (1975). "The Polluter Pays Principle", OECD, Paris.

PCBs Core Group (Bailey, M L, convener), Hazardous Wastes Task Group (March 1988). "A Strategy for Managing PCBs", Ministry for the Environment, Wellington.

SUBSTITUTION OF HAZARDOUS CONSTITUENTS OF HOUSEHOLD PRODUCTS

Kim Christiansen, Technological Institute, Copenhagen
Erik Hansen, Cowiconsult, Copenhagen

Introduktion

Pollution endangering the environment and human health is often a consequence of the use and waste disposal of household products containing hazardous constituents. Hazard is defined as the potential of a chemical to cause harm, and it is a function of both exposure to the chemical and the (potential) toxic effects on living targets.

When introducing new chemical substances to e.g. the EEC member states a hazard evaluation system visualized by a labelling system is enforced based on inherent physico-chemical and toxicological properties. But many daily life products contain chemical substances in small amounts, which involves environmental problems not in the use situation but after use - when the product is turned into waste.

Hazardous waste from households

A compilation of typical household products giving rise to potential hazardous waste is shown in table 1. Sources, amounts and management including suggestions for collection systems are discussed in the international perspective in Yakowitz (1986).

Heavy metals in waste products will when the waste is incinerated be dispersed in the air or landfilled with slags and ashes. Both human and environmental exposure to many metals have reached levels calling for immediate reduction at source. One of the main sources of mercury and cadmium in household waste is batteries but for mercury sources like dentists should also be accounted for. These waste types are easily collectable. Through separate collection of metal wastes the amount of metals in slags and flyash e.g. nickel, chromium, copper, and cadmium in alloys and solders can be reduced. Many electrical appliances e.g. radio & T.V., refrigirators and coffee automates contain contacts, solderings and wires with heavy metals but also insulation foam and cooling systems with CFC, and these waste types are also available for collection. Special notice should be given to plastics e.g. PVC and preserved wood containing heavy metals.

Table 1 List of typical household products considered potential hazardous waste when disposed. (Rewriten from Yakowitz, 1986, Enterprise for Education, 1986, and Hansen, 1988).

Cleaning products	Polishers for furniture and stoves, toilet and sewage cleaners, carpet and upholstry cleaners, disinfection fluids, spot removers, decalcination fluids etc.
Paints	Paints and lacquers based on organic solvents e.g. mineral spirit, water-based paints (glycols), wood preservatives, kerosene a.o. solvents, paint strippers, thinners, joint resins etc.
Maintenance products	Textile preservation, shoe polish leather preservatives, rug cleaner, rust removers and proofing, car lacquers, cooling liquids, antifreeze, engine cleaners, brake fluid, pesticides (insecticides, fungicides, fumigants, herbicides), swimming-pool chemicals etc.
Home work and office	Photographic chemicals, glues and cements, soldering liquids, inks, liquid correction fluids, textile dyes, artist paints and oils, cosmetics etc.
Others	Batteries (cylindrical and buttom cell types e.g. alkaline and mercuryoxide), mercury thermometer, car fuels, medicines and drugs (syringes), vacuum cleaning bags, fluorescent lamps (both tubes and ballast), asbestos materials, tarry materials, aerosol cans (CFC), refrigirators and freezers, autobatteries, rechargable accumulators etc.

Another acute pollution problem from incineration is emissions of dioxins, which not fully can be reduced through better operation control and flue gas cleaning. Separation of chlorine containing wastes e.g. PVC, food wastes and bleached paper may reduce the dioxin generation, but in general these wastes are not easily collectable. The National Agency for Environmental Protection is deeply involved in research in substitution possibilities for PVC, and for the general chlorine question a Danish materials flow analysis is in preparation.

Other types of potential hazardous wastes e.g. most of the organic chemicals as paint and solvents are not assessed as problematic when incinerated in dilute concentrations. On the other hand these waste categories will give work environment problems for waste collection personnel and eventually for the sewage system. Furthermore the educational effect on the public through separate collection and the positive effects on the quality of compost and other recyclables should not be underestimated.

In the following the possibilities for and problems with substitution of cadmium (Christiansen et al., 1988) and CFC (Hansen and Thompsen, 1988) are given.

Cadmium in plastics

In the first months of 1988 a proposal from the Commission of the European Communities for an action plan on cadmium was adopted by the Council of Ministers (CEC, 1987). The action plan includes monitoring of cadmium pollution, limitations to the discharge of cadmium to air, soil and water, and incentives to develop alternatives to plating and pigmentation and stabilisation of plastics with cadmium and other use restrictions. Most non-essential uses of cadmium in plastics will be banned.

The background for this action plan is national legislation restricting cadmium use already in force in Denmark and proposed in the Netherlands. Also Sweden and Switzerland have enforced use limits. Behind the political reasoning is the fact that the WHO provisional tolerable weekly intake of 400-500 ug cadmium, which has been widely adopted, has already been reached in some countries. In Denmark cadmium intake with food and smoking and through exposure in work environment etc. is expected to cause kidney damages in 1000 to 30000 Danish citizens at the age of 45-65.

Cadmium is an inevitable by-product to zinc production. World production is approximately 15000 tonnes and is used for plating (25%), batteries (35%), pigmentation (20% - of this 80% in plastics), plastic stabilizing (15%) and alloys (5%) - in round figures. Cadmium pigments used in bulk plastics i.e. polyethylene, polypropylene, PVC, ABS a.o. accounts for 85% and technical plastics e.g. polyamide, polycarbonate and PETP for 15%. In Denmark main uses through import and own production are in cars, packaging, toys, furniture and kitchen articles. The typical cadmium concentration in a plastic product is 0,1% - 1,0%.
Cadmium is also used for stabilisation of mainly PVC as protection against degradation from ultraviolet light. Dominating are window frames and door frames for outdoor uses with minimum 2200 ppm cadmium but indoor uses in homes, cars etc. in concentrations around 500 ppm are not uncommon.

Cadmium pigments can be substituted with other inorganic pigments e.g. based on lead, iron, titanium and chromium and combinations or with organic pigments (insoluble) or dyes (soluble) for most uses. Main disadvantage is the loss of brightness and durability of the colour. Another problem is that only cadmium is stable at 500-600°C, which is a typical process temperature for some technical plastics. Alternatives exist but they are very expensive.
The substitution enforced by national legislation in Sweden (and in Denmark) have risked work environment problems due to the very dusty character of organic pigments and dyes. Also long term health effects are not known as they are for most of the inorganic pigments. Incineration af plastic with other inorganics will in the end lead to dumping of potential hazardous waste with slags and flyashes.

Development of co-stabilizers e.g. barium-zinc systems have given realistic alternatives to the barium-cadmium stabilizers used for soft and semi-hard PVC. For indoor uses it is not necessary to use cadmium as illustrated by several European product examples. Also for outdoor uses alternatives exist but substitution is limited especially for building articles due to standardized demands to durability of 10-15 years.
Danish regulations have limited the use of cadmium pigments to 1-2 years with the exception of security colouring of e.g. natural gas pipes for another 5-10 years. Indoor use of cadmium-stabilized PVC is now forbidden. For outdoor uses window and door frames are allowed with cadmium for another 5-10 years giving time to technical developments and field testing.

As in Sweden the Danish ban does not include uses in the military and transport sectors. The amount of cadmium used in these sectors is not known. Swedish authorities have estimated a use reduction in Sweden from 100 tonnes per year to 50 tonnes over 10 years, and the same relative reduction is acknowledged for the use of cadmium-pigments in Europe over 5 years. Danish and other countries legislation will increase this reduction.

Chlorofluorocarbons

CFC are assumed to have serious effects on the ozone layer of the stratosphere, and this is expected to cause severe effects on men and the environment due to increased radiation etc. To avoid this a large number of nations in September 1987 agreed with the Montreal-protocol to reduce the use of CFC from first of January 1989.
According to the protocol all nations will have to cut their CFC-consumption
- to the 1986-level from 1989-1990
- to 80% of this from 1993-94
- and to 50% of the 1986-level from 1998-99.

It should be noticed, that the Nordic Countries, especially Norway and Sweden, intends to enforce more rapid reductions in the CFC-consumption.

CFC are used in a large number of products of which some are of great importance in buildings and in households. An overview is given in table 2.

Table 2 Typical uses of CFC in buildings and households and means of substitution.

Products with CFC of which the major part will be released at disposal	CFC-function	Means of substitution
Building insulation including ground insulation in doors and gates made of urethane or styrene foam	Foaming and insulation agent	New fluorcarbon compounds. Other insulation material. Other foaming agents
Domestic refrigerators and freezers	Foaming and insulation agent of the urethane. Heat transmission	New fluorcarbon compounds
Heating pumps	Heat transmission	New fluorcarbon compounds. Improved construction
Surfboards, toys etc. containing urethane foam	Foaming agent	Other foaming agents or constructions

Products from which CFC are released during production or by use		
Mattresses, furniture etc. made of flexible urethane foam	Foaming agent	Other foaming agents
Aerosols, paints, hairsprays etc.	Propellant	Other propellants (buthane, carbondioxide etc.)
Shoesoles, bicycle saddles car parts, flower blocks made of rigid urethane foam	Foaming agent	Other foaming agents

The means of substitution and other methods to reduce
emissions of CFC has been evaluated in a Nordic investigation
(Hansen and Thomsen, 1988). Based on this evaluation table 2
also indicates in general terms promising means of substitution.
Regarding aerosols, which accounts for app. 50% of the total
consumption in EEC in 1983-figures (EEC, 1987b), use of CFC
has already been restricted in Denmark, Norway and Sweden.
For most uses CFC are substituted with buthane, propane or
carbondioxide. In Denmark the regulation is expected to reduce
the CFC-consumption from approximately 730 tonnes in 1986 to
approximately 100-150 tonnes per year (Hansen and Thomsen,
1988).
For other uses substitution is a choice between other chemical
compounds or other materials and constructions. In products
made of urethane foam CFC use may be reduced through increased
use of di-isocyanates, which by reaction with water forms
carbon dioxide as foaming agent. Products manufactured this
way will have larger densities and lesser insulation effect,
and also they will be more expensive. Nevertheless, substitution to water-carbondioxide may be a realistic alternative for
many building insulation purposes. Other alternatives are
mineral wool etc.

Regarding domestic refrigerators and freezers, the best alternatives to CFC seems to be the new fluorcarbon compounds e.g.
CFC 134a, CFC 141B etc. These compounds are expected to be
commercially available in 1992-96 pending on toxicity testing
and process development turning out succesfully (Dupont,
1988).

Conclusion

The examples of cadmium and CFC show that substitution is
possible for many uses of these hazardous constituents of many
household products -at least from a technical point of view.
This seems to be the case for many other hazardous
constituents, but substitution is often limited by economic
considerations and doubts about product quality. Conservatism
in the selection of raw materials and processes and lack of
knowledge of alternatives are other points to be made.
Danish authorities and industries are gaining valuable experiences these years with substitution possibilities for cadmium,
mercury, PVC and CFC as the main priorities. The un-assessed
negative effects on human health and environment of
substitutes for cadmium in plastics - due to lack of knowledge
- has now focused interest and funding to this key issue of
any substitution.

References

Christiansen, K., Hansen, E. and Svanum, M.(1988): (Substitution possibilities for cadmium in plastics). In Danish. Summary report to the National Agency for Environmental Protection. Technological Institute, Copenhagen. (To be published).

CEC(1987): Environmental Pollution by Cadmium. Proposed Action Programme. Commission of the European Communities, Bruxelles.

DuPont(1988): Fluorcarbon/Ozone-Update. Wilmington, USA.

EEC(1987): The State of the Environment in the European Community 1986, Luxembourg.

Enterprice for Education, Inc.(1986): Hazardous Wastes from Homes. Santa Monica, USA.

Hansen, E.(1988): (The use and disposal of household chemicals in Denmark). In Danish. Report to the National Agency for Environmental Protection. Cowiconsult, Copenhagen. (In press).

Hansen, E. and Thomsen, H.(1988): Reduction of CFC-consumption. In Danish and English. Report to the Nordic Council and the Council of Nordic Industrial Federations. Cowiconsult, Copenhagen. (To be published).

Yakowitz, H.(1986): Fate of Small Quantities of Hazardous Waste. Environment Monographs No. 6. OECD, Paris.

SAFE, ZERO-IMMISSION ULTIMATE DISPOSAL OF SOLID HAZARDOUS WASTES IN SALT CAVERNS

Dipl.-Ing. F. Crotogino*
Dr.-Ing. Dipl.-Geol. H. J. Schneider*

*) KAVERNEN BAU- UND BETRIEBS-GMBH, Roscherstraße 7, 3000 Hannover 1, Germany,
Tel.: 0511/34 846-0

0.   INTRODUCTION

Waste management is currently characterized by two trends, the reduction in quantity of "traditional" waste products as a consequence of more processing and incineration and an increasing number of new wastes. These new categories of waste material - predominantly residues from flue gas cleaning plants at waste incineration facilities - bring their own management problems. They are extremly difficult to cope with because of the, in general, large amounts of water soluble constituents.

The ultimate disposal of these wastes in dumps above ground cannot be recommended as it may create a new waste treatment problem.

It is preferential to dispose of all waste materials, which are currently not suited to management using reprocessing or neutralization, in impermeable geological formations underground.

Solution-mined salt cavern repositories offer an acceptable technical and economic method for the ultimate disposal of this category of solid hazardous waste, which can not be further decontaminated and reduced.

This paper presents the technical concept of salt cavern repositories, discusses the geotechnical and processing aspects and indicates the geological prerequisites.

1. SAFETY ASPECTS OF DISPOSAL FACILITIES

The most important requirement for waste disposal facilities is now recognized to be the complete and hermetic enclosure of the toxic materials against the biosphere, i.e. not even the release of diluted toxic substances can be permitted to occur, since neither the toxicological effects of the individual substances nor the synergism of the toxic substances on the biosphere can be clearly predicted.

Current standards for the containment of toxic matter in waste disposal facilities are based on multi-barrier systems, in which, as far as possible, the individual barriers are redundant and can be monitored [1]. However, the safety of the waste disposal facility is not only determined by the barrier system itself, but also by the behaviour of the waste substances within as well as the ambient conditions of the disposal facility.

Multi-barriers, such as the basement seal, the seepage water drainage and surface seal, as applied in standard landfill disposal facilities, are passive barriers designed to retain and screen toxic materials.

Technical and geological barriers are very different and should be considered separately. It is generally assumed that technical barriers at waste facilities only have a limited lifetime. In comparison certain geological barriers are capable of ensuring that no toxic substances are released into the biosphere over geological time periods. Geological barriers of this type are suitable for the final disposal of radio-active waste. Salt, due to its favourable petrophysical properties, such as its impermeability to water and gas, is particularly well suited as a host rock for the entombment of toxic waste, as in the case of the Herfa-Neurode potash mine in West Germany [2].

It is common knowledge that most commercially available liners for basement sealing are permeable to chlorinated hydrocarbons and organic solvents. For this reason waste disposal facilities may not simply be regarded as sites for the indiscriminate dumping of waste, as standard household waste landfills have shown. Indeed, the controlled disposal and storage of waste products, i.e. the sorting of waste substances for a particular type of waste disposal facility and the pretreatment of waste materials by chemico-physical processes, must be introduced to enable prognosis of the long-term behaviour of the waste products in the disposal facility.

Another significant aspect for waste disposal facility safety is the influence of nature on the facility itself, such as weathering, frost, precipitation, erosion and other geodynamic processes. Here, too, subsurface repositories are clearly superior to surface waste disposal systems.

These safety aspects therefore make it essential that the function of the waste disposal system as a final repository or interim storage be clearly defined to enable future generations to estimate and predict the hazard potential of a waste disposal facility. Accordingly, a final repository must fulfil what the name implies: the final disposal of waste; the toxic substances must be sealed from the biosphere for geological periods. The repository must be inaccessible, the disposed waste non-recoverable. Moreover the facility must be maintenance-free and not require monitoring.

Waste products are only stored for limited periods in interim storages. At expiry of the storage period the waste products are recycled or reprocessed. The storage must therefore be easily accessible and the waste materials be easily recoverable. The technical barriers require constant monitoring and maintenance during the operational life-time.

2. POSITION OF WASTE DISPOSAL FACILITIES IN WASTE MANAGEMENT

In traditional waste management the waste disposal facility is regarded as the end of the line, where everything is delivered and dumped. Waste disposal facility safety requires thorough rethinking, especially with regard to long-term prognosis of waste products in waste disposal facilities. To make this possible, it is necessary to restrict the waste spectrum intended for disposal, with associated preselection and treatment and/or detoxification and possibly volume reduction.

This requires a reorganization of overall waste management policy. This has been done in the Federal Republic of Germany in the 4th Amendment to the Law on Waste [3]. Accordingly, priority is placed on the avoidance and recycling of waste products. Where this is not possible, waste products shall be treated by chemical or physical processes to reduce the toxic and hazard potential or, if combustible, be incinerated for volume reduction.

Environmentally compatible residues are recycled. Other residues are stored for grading into waste flow paths, that is, waste with foreseeable recycling potential is diverted to interim storage, and the rest to a final disposal facility, where these highly toxic, easily soluble, leachable wastes are again graded according to waste type and disposed of in salt mines or salt caverns.

3. WASTE SPECTRUM FOR SALT CAVERN REPOSITORIES

Pursuant to this waste management concept, the State of Lower Saxony has adopted the policy of the disposal of essentially highly soluble largescale waste in salt caverns [4]. These wastes are mainly ashes, slag and dust from waste incineration plant, mineral sludges, salts, and catalysts. Salt caverns represent a particularly important waste disposal means for the group of residues produced by waste incineration, because it is anticipated that the incineration of household and hazardous waste will greatly increase in West Germany in the immediate future.

The analysis of the solubles of dissolved dust from waste incineration (Tab. 1) shows that these consist mainly of calcium, sodium, chlorides and sulphates as well as a number of heavy metals in low concentrations. As there is no recycling potential for these materials, the only alternative is final disposal in deep salt formations.

	mg/l
Calcium	1 023,0
Magnesium	0,3
Sodium	803,0
Zinc	0,19
Cadmium	0,03
Mercury	0,002
Thallium	0,002
Molybdenum	0,26
Arsenic	0,001
Chloride	2 800,0
Fluoride	7,04
Sulfate	2 244,0

Tab. 1: *Eluate Analyses of Dust from Waste Incineration Plants*

4. SALT CAVERN REPOSITORY TECHNOLOGY CONCEPT

The engineering concept of salt cavern repositories meets all requirements placed on waste disposal systems in the future. It serves as a final disposal system for non-recyclable waste [5]. The entire facility consists of only the subsurface cavern section, the actual repository, and the limited surface installations (Fig. 1).

A salt cavern repository is a multi-barrier system. The barriers are natural geological barriers in deep salt formations, which, on account of their impermeability and their distance from aquifers, ensure an effective seal against the biosphere.

Fig. 1: Salt Cavern Repository

Salt caverns are cavities constructed in salt rock by solution mining [6].

After completion of the solution mining process the caverns are pumped free of brine. Once the brine has been completely removed the process of waste emplacement starts. This requires the treatment of the waste in the surface plant to prevent reactions between the different waste products and to process the waste products to meet the requirements for the filling procedure.

The treatment system is based on four treatment branches, in which solids are crushed and selected depending on grain size, dusts are agglomerated, and sludges and slurries neutralized and dewatered. The four treatment branches converge for forming the treated waste into pellets or brickets.

The pellets and brickets are then fed into the cavern via a free-fall pipe. After the cavern has been completely filled with waste, it is permanently sealed against the biosphere. The uncased well section is packed with salt and the cased section filled with a bitumen-cement mix. The surface facilities are removed, and the site is recultivated and returned to its original use. Long-term monitoring and maintenance are unnecessary because the waste is sealed from the environment for geological periods by the geological barriers of the salt rock and the well seal.

As a consequence of the complete back-filling of the cavern there are no follow-up problems from cavern stability and convergence. The waste products in the cavern are turned to a monolithic salt body in the long-term under the pressure and temperature conditions prevalent in the salt rock mass.

5. CONCLUSION

The creation of hazardous waste problems for future generations by the use of salt caverns can be safely disregarded. Moreover, follow-up costs do not arise with this type of waste repository system.

However, salt cavern repositories should not be regarded as the cure to all evils of the waste industry. Rather, salt cavern waste disposal sites should be seen as one component within an overall waste management concept for certain waste types. The question "to burn or bury?" should not be regarded as representing all alternatives. Indeed the catch phrase should be "burn, treat and bury". Unavoidable waste which cannot be recycled must be treated physico-chemically to reduce the toxic potential and waste volume, and only that residual waste which cannot be recycled should be filled into repositories.

Extremely high safety requirements must be placed on repositories for these residual wastes to keep current and future environmental risks to a minimum. Our legacy to future generations must not be in the form of waste "problem" sites.

REFERENCES

[1] STIEF, K. (1986):
Das Multibarrierenkonzept als Grundlage von Planung, Bau, Betrieb und Nachsorge von Deponien. Müll und Abfall, H. 1, S. 15-20.

[2] UTD:
Untertagedeponie Herfa-Neurode. Firmenbroschüre. Kali und Salz AG, Kassel, S. 16.

[3] Gesetz über die Vermeidung und Entsorgung von Sonderabfällen (Abfg). Bundesgesetzblatt, Jahrgang 1986, Teil 1.

[4] Rahmenplan Sonderabfallbeseitigung Niedersachsen. Der Niedersächsische Minister für Ernährung, Landwirtschaft und Forsten, Hannover, 1985.

[5] SCHNEIDER, H.J. (1986):
Enddeponierung von Sonderabfällen in Salzkavernen. In: Handbuch der Müll- und Abfallbeseitigung, Nr. 8192, S. 1-19, Erich Schmidt Verlag GmbH.

[6] QUAST, P. und BECKEL, S. (1981):
Derzeitiger Stand der soltechnischen Planung von Speicherkavernen im Salz und die damit erzielten praktischen Ergebnisse. Erdöl-Erdgas-Zeitschrift, H. 6, Jg. 97, S. 213-217.

CASE STUDY OF THE 250.000 m³ ACID TAR DISPOSAL PIT ABOVE THE RIJEKA PORT

M. Ivanc, B. Pleskovič, SMELT Ljubljana, Yugoslavia

The acid tar disposal pit above the Rijeka port, containing over 250.000 m³ of hazardous waste, has been used by the Rijeka Oil Refinery and other industries since 1950. It is situated in permeable limestone terrain, 314 m above the sea-level and 4 to 5 km from the seaside wells lying in its hydrogeologic influential zone. This pit, called Sovjak pit, represents a constant environmental danger, causing the pollution of air and water resources.

The main groups of refuse disposed into the Sovjak pit in the period 1979 - 1981 were the following in percents by weight: 29.4% acid tars, 8% heavy tars, 8.4% tank slops, 5.4% heavy oils, 7% oil emulsions and 41.8% carbide sludges from acetylene production in shipyards. The total quantity was approx. 20.000 t/year.

In 1984, the company Smelt, Ljubljana, began to elaborate a complex study of the impact on the environment in order to find the best solution for successive decreasing, closing, and recultivating this pollution source. Further dumping of acid tars and aggressive or toxic refuse has been forbidden, while dumping of alkaline carbide sludges and other alkaline refuse remained allowed, though in dehydrated condition only. Thus, the compositions of refuse disposed in the 1985 - 1987 period into the Sovjak pit changed to the following shares in percents by weight:

65% carbide sludges, 20% oil emulsions and tank slops, 15% tars and oil fractions. The quantities decreased to 7,250 t in 1985, 6,492 t in 1986 and 4,280 t in 1987.

In the meantime, more and more industries have changed and adapted their production, in order to minimize or avoid the quantity of refuse inadequate respectively not any more allowed for the Sovjak pit, and/or introduced alternative disposal methods. In 1986, the Oil Refinery finally opened its incineration plant of the fluidized bed type and developed solidification techniques for acids and other problematic refuse, unsuitable for incineration. In this way, the necessary prerequisites for a definitive closing and recultivating of the Sovjak pit have been created.

The main goal of the mentioned complex study, carried out by Smelt, was to estimate the cumulative impact of the Sovjak pit on the environment in order to determine the optimal remedial measures and the most suitable technology for closing and recultivating of this pit. An insight into the composition of refuse within the pit and into physical and chemical changes with depth was necessary, which required an adequate sampling. The extraction of bore samples was very difficult because of **the soft and unsuitable pit surface and** due to adverse and toxic gas emissions. A special pontoon for heavy drilling rig was inevitable as well as protective respirators.

The main analytical data on the centrally located bore samples are given in Table 1. The surface crust of soft tar approx. 1m thick was followed by a 2.3 m thick water layer. This water layer, which penetrates the surroundings, contais a rather normal pH value (6.9), what is surely the result of alkaline carbide sludges from the shipyards. The heavy metals are mostly bound with the thicker layers and contained only in traces in the water layer. The density of refuse increased very quickly with depth, so the samples below 9 m can be specified as solid matter. Due to the increased density and rigidity, the drilling below 9 m progressed very slowly and was practically impossible below 14 m. The permeability of refuse below 12 m rapidly decreases from 10^{-7} cm/s to 10^{-8} cm/s. These low permeability coefficients prove that the refuse below 12 m depth is aggregated and physically integrated into a thick asphalt-like barrier which prevents vertical leakage of hazardous chemicals into the surroundings.

The water balance analysis of the Sovjak pit showed that approx. 11,400 m³/year of contaminated water escape through rock cracks not yet sealed by the heavy tar, up to 3 m depth. The comparison of pollution data on wells inside and outside the Sovjak pit influential zone, using the existing reports for the 1975 - 1987 period, rendered possible an estimation of the cumulative pollution. Only the Zvir 2 well is under the exclusive influence of the Sovjak pit while the other three mentioned wells are subject to additional pollution from the upper city areas. Nevertheless, all pollution parameters of the four wells lay safely below the allowed limits while the Zvir 2 well does not give any evidence of a greater pollution than in the case of wells situated outside the Sovjak pit influence zone. The relevant comparative pollution data are presented in Table 2. A similar comparison of the heavy metals concentrations for all mentioned wells proved that no significant difference could be found and that these concentrations lay far below the limits allowed, including the concentrations of Pb and Cr, which have been relatively high in the water layer of the Sovjak pit.

The main conclusions of this case study are as follows: (i) No significant pollution of the seaside wells could be established). (ii) The approx. 2 m thick water layer under the surface crust is the most mobile part of the pit content. The lower layers are rather dense and physically integrated into a thick asphalt-like barrier which prevents most of vertical leakage. (iii) Since only alkaline carbide sludges are accepted, the mentioned water layer is practically neutral and relatively little polluted. (iv) A forced solidification of refuse requires complicated excavations and outside treatment, which is capital intensive. The Sovjak pit can be also closed gradually by dumping of dehydrated inorganic alkaline production waste, followed by a final recultivation. (v) Other refuse not any more acceptable for the Sovjak pit is now being incinerated or solidified within the new facilities in Rijeka.

CASE STUDY

Table 1. MAIN ANALYTICAL DATA OF THE BORE SAMPLES FROM THE SOVJAK PIT

Number and Name of Sample	Depth m	Water Content g/kg	Neutr. No. mg KOH/g	Total Sulfur g/kg	pH	Heat Value MJ/kg	Glow Rest W %
1. Tar, soft	surface	150	4.3	22.2	4.0	38.15	–
2. Tar, soft	0 – 1	120	4.6	21.8	3.8	42.92	0.39
3. Water	1 – 3	–	1.2	–	6.9	–	–
4. Water	3 – 3.3	–	2.0	–	6.3	–	–
5. Tar, heavy	3.3 – 4	235	16.4	38.4	2.6	21.83	12.0
6. Tar, heavy	4 – 4.5	304	15.7	40.3	2.4	25.72	12.4
7. Tar, dense	4.5 – 5	282	14.0	39.2	2.5	26.46	12.8
8. Tar, dense	5 – 6	270	14.7	41.7	2.5	25.25	13.5
9. Tar, dense	6 – 6.5	264	14.9	39.6	2.5	26.46	13.2
10. Tar, dense	6.5 – 7.1	272	13.8	43.7	2.4	24.28	–
11. Tar, dense	7.1 – 8	276	15.2	39.4	2.5	25.00	13.7
12. Tar, dense	8 – 9	348	–	38.7	–	24.65	16.1
13. Tar	9	–	98.8	61.0	–	21.75	9.0
14. Tar, semi hard	9 – 10	–	110.3	56.3	–	19.29	23.0
15. Tar, semi hard	10	–	–	44.0	–	21.30	–
16. Tar, hard	10.2 – 11.4	–	–	61.0	–	18.65	24.7
17. Tar, hard	11.4 – 12	–	–	70.4	–	19.24	15.4
18. Tar, very hard	12 – 13	–	146.2	80.8	1.0	25.32	12.9
19. Tar, very hard	13 – 13.5	–	113.8	77.3	1.1	25.74	–
20. Tar, very hard	13.5 – 14	–	125.9	87.1	1.0	25.24	–

Table 2. COMPARATIVE POLLUTION DATA FOR WELLS WITHIN AND OUTSIDE THE SOVJAK PIT INFLUENTIAL ZONE FOR THE 1975 - 1987 PERIOD

Parameter	Date	Outside infl. zone				Within infl. zone		
		Martinščica w. 1	2	3	Zvir 2	Jelšun	Mlaka	3.maj
pH	06.05.75	7.4	7.5	7.5	7.55	7.3	7.4	7.35
	18.05.77	7.45	7.45	7.45	7.5	7.2	7.6	7.2
	04.05.83	7.6	7.6	7.6	7.5	7.3	7.4	7.3
	29.08.83	7.3	7.5	7.3	7.4	7.2	7.3	7.2
	09.01.84	7.4	7.4	7.4	7.4	7.35	7.4	7.35
	03.09.84	7.5	7.6	7.5	7.5	7.2	7.4	7.2
	13.01.86	7.5	7.5	7.5	7.5	7.45	7.4	7.45
	28.08.86	7.5	7.5	7.5	7.5	7.3	7.4	7.4
	05.01.87	7.5	7.5	7.6	7.6	7.2	7.4	7.2
	04.05.87	7.5	7.5	7.5	7.5	7.5	7.4	7.4
Sulfates mg/l	06.05.75	3.0	3.0	3.0	5.0	3.0	6.0	4.0
	18.05.77	4.0	4.0	4.0	5.0	10.0	15.0	10.0
	04.05.83	8.0	8.0	8.0	7.0	10.0	8.0	7.0
	29.08.83	10.0	4.0	4.0	6.0	9.0	15.0	5.5
	09.01.84	7.0	7.0	6.0	10.0	12.0	20.0	15.0
	03.09.84	4.0	2.0	2.0	3.0	10.0	10.0	13.0
	13.01.86	4.0	5.0	8.0	8.0	10.0	15.0	13.0
	28.08.86	5.0	5.0	5.0	6.0	8.0	15.0	13.0
	05.01.87	5.0	6.0	7.0	4.0	10.0	10.0	10.0
	04.05.87	6.0	8.0	8.0	5.0	10.0	12.0	12.0
K Mn O_4 mg/l	06.05.75	4.4	4.4	4.8	7.6	6.0	4.4	3.2
	18.05.77	4.0	3.2	3.2	5.2	6.0	3.2	6.0
	04.05.83	4.0	4.0	2.8	8.4	5.6	5.6	4.8
	29.08.83	3.6	4.4	4.0	4.8	4.4	4.4	4.8
	09.01.84	3.2	4.0	6.8	7.2	3.2	4.0	3.2
	03.09.84	6.0	5.2	5.2	6.4	5.6	5.2	7.2
	13.01.86	2.8	2.4	2.8	4.0	2.8	2.8	3.6
	28.08.86	4.4	4.4	5.6	5.6	8.9	4.4	4.8
	05.01.87	5.2	6.8	4.8	7.6	8.0	6.4	8.4
	04.05.87	6.8	6.6	6.4	5.2	7.2	7.6	6.8

Source: Adapted from the ZZZZ - Rijeka Archives

MANAGING SITES CONTAMINATED BY HAZARDOUS WASTES IN QUEBEC

Claudette Journault and Jean-Pierre Plamondon
Ministère de l'Environnement du Québec

The management of sites contaminated by hazardous wastes or other hazardous substances has been undertaken by the Ministry of Environment of the province of Quebec through two recent programs that address this difficult problem.

1 - THE GERLED PROGRAM

The GERLED program started in October 1983 by Quebec Ministry of Environment had the following mandate:

- to identify and prepare a preliminary assessment of sites that had potentially or actually been used to dispose of hazardous wastes in Quebec and to establish action priorities based on this assessment;

- to prepare and execute a plan of action to remediate contaminated sites;

To this day of the 1095 sites considered, 333 have been identified in the inventory, 66 as category I priority, 102 as category II and 165 as category III.

Assessment work has been undertaken on some 101 sites and remedial action feasibility studies or remedial action has been started on 54 sites.

Up to now, this program has been quite successful in getting the collaboration of many responsible parties (mostly at industry owned sites) without having to resort to regulatory action.

As the GERLED program went on, an increasing number of landowners and developers wishing to sell or reuse former industrial sites sought the advice of the Ministry on the need to do some remediation work on their sites and on the degree of decontamination to be achieved before development. The second part of the Ministry's action on contaminated sites tries to bring an answer to these questions.

2 - THE CONTAMINATED LAND REHABILITATION POLICY

The phenomenon of closing and dismantling former industrial sites will be increasingly prevalent in years to come. Since these industrial sites are often well located and affordable, they are eagerly sought by promoters wishing to reuse them for other purposes.

Aware of the fact that industries may have contaminated the soil and groundwater while producing or utilizing hazardous substances, the Ministry of Environnement, after consultations in some countries that were already faced with similar problems, has established (a) key criteria indicative of soil and groundwater contamination, (b) a list of activities that may have resulted in site contamination, and (c) a procedure for effective management of contaminated sites to ensure that the use of former industrial sites does not threaten public health or the environment and to encourage maximization of urban potential.

This contaminated land rehabilitation policy, made public in March 88, can be summarized as follows:

- it is preferable to permit the recovery of contaminated sites provided that the quality of soil and groundwater is compatible with planned uses;

- a promoter planning to reuse a former industrial or any other potentially contaminated site shall first conduct a site study to assess the nature, extent, and current or anticipated impacts of contamination;

- before it is reused, a site contaminated by hazardous substances shall be decontaminated in accordance with the key criteria established by the Ministry and with the anticipated uses.

- In order to ensure safe decontamination and provide for permanent solutions contaminated soil shall be handled treated or disposed of through methods that are environmentaly acceptable.

- studies and work associated with assessment and rehabilitation of contaminated soil and groundwater shall be carried out by either the owner or the promoter.

a - **Key criteria**

These criteria, that cannot be presented in full here because of space restrictions, have been determined by the Ministry both for soil and for groundwater for 26 inorganic and 52 organic parameters (including pesticides). They have been published in the form of a table giving 3 sets of values (A, B & C) that define different ranges of concentration requesting different levels of actions.

Inspired by the Dutch model these criteria have been expanded and adapted to Quebec's conditions and background. They should not be considered as legal standards: they are to be used by the specialists that have to assess the contaminated sites, as guidelines, taking into account site-specific conditions and every project's particularities. In that respect, the Ministry is also ready to consider the validity of site specific criteria established after a detailed risk analysis study.

b - List of activities likely to contaminate soil

The list that has been prepared by the Ministry is the following.

Land disposal of waste and other residue:

- sanitary landfill, dry materials and garbage dumps, industrial wastes dumps, snow disposal sites, mining residue disposal sites.

Industrial and commercial activities:

- chemical and petrochemical, pharmaceutical, pesticide, fertilizer, paint and lacquer, pulp and paper, metallurgical, electrotechnical, galvanizing, wood preservation and textile industries; shipyards, foundries, coking plants; battery, used oil, solvent, liquid waste and barrel recycling plants; service stations, dry cleaners, repair and maintenance workshops for bus and subway, automobile as other vehicle; transformer substations.

Storage and transfer of hazardous substances:

- storage of chemical and petrochemical products, pesticides, solvents, oil pipeline corridors.

Land spreading of sludges and liquids:

- land spreading of contaminated sediments, of water treatment or septage sludge and land spreading or land farming of petroleum residues.

c - <u>Procedure for effective management of contaminated sites</u>

This procedure defines the role of the major parties whose collaboration is essential to the success of the policy: the landowner or developer, the municipality and the Ministry of Environment. It can be summarized as follows.

The owner or developer warned by the municipality (or any other informed party) of the contamination potential of his site should, before going on with his transaction or project, obtain the opinion of the Ministry as to the compatibility of his project with the principles exposed in the policy.

To enable the Ministry to do so, the developer must realise a site assessment study and, if contaminants are actually present at an unacceptable level, he must submit a proposal for site rehabilitation.

The owner or promoter always remains responsible for the environmental impacts of his development project, the municipality has to notify the promoter and the Ministry of possible problems as development projects occur and the Ministry provides technical support assisting the responsible parties in making proper decisions.

A policy based on awareness

The contaminated land rehabilitation policy was drafted to solve a new type of environmental problem that industrialized societies have only just begun to face.

Since the effectiveness of this policy depends, to a large extent, on the climate of cooperation to be established between the parties, the Ministry chose to base the rehabilitation policy on information rather than opt for coercive measures.

In so doing, the Ministry hopes that all will understand it is in the interest of citizens, future users and promoters of a development project to prevent the negative impacts that may result from contaminated sites.

INVOLVEMENT OF MUNICIPAL GOVERNMENT IN APPROPRIATE OR
ENVIRONMENTALLY ACCEPTABLE TREATMENT AND DISPOSAL OF
HAZARDOUS OR TOXIC INDUSTRIAL WASTES IN OSAKA

KIYOSHI KAWAMURA, CHIEF STAFF
INDUSTRIAL WASTE GUIDANCE DEPT., SERVICE DIVISION
THE PUBLIC CLEANSING BUREAU OF OSAKA MUNICIAPL GOVERNMENT

1. Introduction. Along with the active industrial development and the prosperous living and working activities by a large population in the urban areas **reflecting high** growth expansion of the Japanese national economy, the public nuisance problem had become a large social issue in 1960s, which caused a serious threat to the public health and environmental quality by waste water and air pollutants discharged from the factories.
The whole society began to question the way of thinking on a economy-dominant basis and asked itself what the affluence is really to the citizens. The people tended to pay more intense attention to the pollution-free comfortable living environment or climate.
In those social circumstances, the public nuisance legislation came into being in 1970, including the amendments to "Basic Law of Public Nuisance Measures", "Water Pollution Control Law", "Air Pollution Control Law" and "Waste Disposal and Public Cleansing Law". In these laws, the basic principle of making the polluters account for the loss or damage incurred, was clearly defined and the social structure for the public nuisance prevention and the environmental protection was thus established.
In order to prevent water quality degradation, air pollution and other public nuisances, the pollutants-emitting plants and factories should install the facilities necessary to treat the waste discharge water and collect the dust and remove any other pollutants. However, these improved pollution control measures could possibly generate the wastes such as the mud or sludge containing hexda chrome or cyanogen and the dusts with hazardous heavy metals.
As and when these wastes **are** dumped in the natural environment without serious consideration, they will again **and eventually cause pollution to** the Environment. Complete solution of water quality and air pollution problems should require the appropriate or environmentally safe treatment of the wastes with hazardous or toxic materials.

2. Measures for hazardous or toxic industrial wastes in Osaka City In the areas like Osaka City which saw rapid urbanization movements going on in rather limited city district, the appropriate treatment of these industrial wastes had been a serious problem which required urgent solutions. Most of the industrial establishments discharging the wastes were small-sized or **medium-sized** factories, particularly in the electro plating industry or printing and photoengraving industry. Electro plating is one of the industries with local dominance in Osaka area and nearly as many as 500 factories in this industry produced about 9,000 tons of hazardous or toxic mud and sludge every year.

These industrial establishments could hardly be afforded the land lots to build the necessary waste treatment and disposal facilities and their financial resources for the purpose were also very limited. For another reason that there were very few business **constructors specializing in** the treatment of these hazardous or toxic wastes, these factories had to leave the untreated industrial wastes to open air storage within the premises of the factories or to illegal dumping. Such situation was extremely serious in terms of the environmental protection and preservation.

The role of the administration or local municipal government could not rest on simply accusing these industrial establishments of their social accountability, which might otherwise have resulted in promoting improper waste treatment or illegal dumping in a secret manner. It was rather more constructive and meaningful for environmental protection that the administration should participate in the building of the industrial waste treatment and disposal facilities by asking the polluting factories to contribute necessary costs in aid of the construction of such facilities, and should promote appropriate or environmentally acceptable waste treatment.

3. Construction of Clean Osaka Center
(1) Background situation for this project
In Osaka City where residential zones are intermingled with industrial complexes of many small-sized or medium sized factories, there were existing and expected many difficulties in fund-raising and land acquisition for each industrial establishment to conduct the intermediate waste treatment individually to make their discharged wastes harmless to the environment. In December 1975, Electro Plating Industry Association of Osaka lodged a plea with the municipal authorities to take a proper action on the

prevailing situation, being followed by many similar representations or appeals from other various organizations or circles.

As far back as April 1975 Council for Environmental Pollution Control of Osaka City made recommendations on "the environmental pollution control considerations concerning the landfill disposal of the industrial wastes at the HOKKO landfill site of Osaka Port", and submitted to the municipal government the practical administrative guidelines. These guidelines deal with the treatment and disposal of hazardous or toxic industrial wastes including mud and sludge and other pollutants discharged from the establishments of the electro plating industry and other six designated industries, and stipulate the provisions for preliminary treatment of the industrial wastes to make them harmless and neutral to the environment and for limiting the disposal of such harmless and neutralized wastes only at authorized landfill sites and no other places.

In compliance with these recommendations and submitted guidelines, Osaka Municipal Government started in November 1976 the construction of the facilities for harmless and neutralized treatment of industrial wastes in order to establish the hazardous or toxic industrial waste treatment and disposal system which is socially accountable with the involvement or participation of the public administration. The facilities were completed in April 1977 and Osaka Industrial Waste Management Public Corporation was formed to operate and manage the facilities.

(2) Name and other particulars of the treatment facilities
Name : Clean Osaka Center
(or COC for abbreviation).
Address: 9-9, Tsuneyoshi 2-chome, Konohana-ku, Osaka City
Area : 3,756.81 m^2
Building structure:
 Steel-frame slate structure with a total floor area of 700 m^2.
Administrative office:
 Two-storied steel-frame mortared structure with a total floor area of 100.83 m^2.
Facilities installed:
 One set of concrete solidification or cementing facilities
 Six waste collection trucks
 Two fork lifts
 Two shovel cars
 One set of ship loading facilities

Start of operation: May 16, 1977

(3) Industrial wastes to be treated at COC
Industrial wastes to be treated at COC include the mud or sludge discharged from the factories in the electro plating, chemicals and other industries after waste water treatment, and heavy metal containing cinders and embers, dusts, and slags, which should comply with the standards for waste receiving at the HOKKO **Landfill site as recommended** and submitted from Council for Environmental Pollution Control of Osaka City.
However, close investigation **is made for the solidification** treatment, and any wastes which are found as inappropriate for cementing are to be eliminated beforehand .
As far as the industrial wastes discharged from the chemical fertilizer manufacturing industry and other six designated industries are concerned, they should be subjected to the solidification treatment even though they meet the requirements in the verification standards by effluent tests as provided for in the "Waste Disposal and Public Cleansing Law".

(4) Treatment capacity
Maximum treatment capacity is 40 tons per day.

(5) Treatment charge
Treatment charge is 32,000 yen per ton of wastes, or 6,700 yen per one drum can (or 210 kg containment vessel as designated by COC).

(6) Scope of operations by COC
 1) Approval
COC examines the contents of the written applications filed by the pollutant-producing factories and **discusses** them with Osaka Municipal Government for review and final approval.
 2) Waste collection
Waste collection is made by specially designed completely encased collection vehicles owned and operated by COC, which go around the transit stations (waste dump junctions installed jointly by waste-discharging industrial establishments) or individual plants or factories for regular pick-up.
 3) **Waste solidification**
Collected wastes are mixed with the prescribed quantity of special cement and chemical agents with chemical and physical functions to seize the heavy metal contents. The mixture is to be blended and kneaded adequately by

using two mixing machines before being put into the concrete solidified form.
4) Concrete solidified mixtures are laid for curing for some time, and then shipped by carrier vessels from pubic piers to the HOKKO Landfill site for final disposal.
5) Test and inspection
COC conducts appropriate tests with the solidification treatment of the wastes and makes inspection to verify whether the concrete solidified materials are harmless or neutralized to the environment.
6) Actual collection data
Annual quantity of the collected wastes amounts to about 8,500 tons.
7) Number of approved industrial establishments
475 establishments in the electro plating industry, and 107 establishments in other industries.

(7) Safety check of concrete solidified materials
1) Pollutant check
As and when any new application is filed with COC for waste treatment, COC will conduct the review of the written application as well as the solidification experiments with the samples of pollutants submitted from the industrial establishment applying for approval. In case that COC treatment process and method should verify that the material can be made harmless and neutralized to the environment, then the approval will be given to the applicant. As and when the experiments indicate that the wastes in question can hardly be treated by the presently applied treatment process or method, then COC shall inform the applicant of that effect and may suggest the need of appropriate preliminary treatment or of the measures to improve the applicant's existing pollutant emission processes in a manner to comply with the treatment at COC. When this suggestion is realized or satisfied, COC will give an approval to the application.
2) Concrete solidified material check at COC plant
With respect to the wastes which are concrete solidified and made harmless and neutralized to the environment, four samples are selected at random every month and they are placed under examination to confirm the safety of the materials concrete solidified at COC.
3) Check at HOKKO Landfill site
With respect to the concrete solidified mixtures delivered to the HOKKO Landfill site, four mixtures are selected at random every month, and they are placed under examination to confirm the safety of such mixtures as and when placed in the natural environment.

4) Analysis results
60 concrete solidified mixtures within COC plant and 48 mixtures placed at HOKKO Landfill site were examined in 1986, and the analysis was made for six items of mercury, cyanogen, hexad chrome, cadmium, lead and arsenic, respectively by a public analysis institute and a private organization. Results submitted respectively by these public and private institutes were cross-examined and it is found that all the tested concrete solidified mixtures have **satisfied** the requirements in the HOKKO Landfill **site receiving standards of which** safety is set three times as stringent as Verification Standards as provided for in the Law.

(8) Future issues
Ten years have passed since 1977 when COC started its operation.
In the meantime, the administration could keep a close watch on the flow of the hazardous or toxic industrial wastes generated in the city areas, beginning with collection, transportation, intermediate treatment and to the final disposal. And the business operation arrangement by COC could play a significant and important role in the appropriate and environmentally-acceptable treatment of the hazardous or toxic industrial wastes, especially by confining them into the limited landfill site and preventing hazardous pollutants from emitting widely into the environment.
However, this waste treatment business has not proved as yet cost-effective and the facilities are exposed to obsolescence. This will pose new issues to be solved.
This is also the time to reexamine the scope of COC business, not only simply continuing the landfill work by harmless and neutralizing treatment of hazardous materials with heavy metals but also developing a new system to recycle **useable** contents of the wastes and reduce the absolute volume or quantity of the wastes as ultimately disposed.

COAL MEDIATED vs. THERMAL HYDROGENOLYSIS OF CHLORINATED WASTES

Robert Louw, Jeffrey A. Manion and Jan-Peter Born
Center for Chemistry and the Environment, CCE,
Gorlaeus Laboratories, Leiden University,
P.O. Box 9502, 2300 RA   Leiden, Netherlands.

Introduction

Organic chlorine compounds have found widespread use for example as solvents, monomers or pesticides. Proper disposal of the wastes from their manufacture, or of disqualified products such as PCB's, is associated with environmental and socio-political problems. Sea dumping or land filling being totally unacceptable now, incineration - whether on land or at sea - is the only practical alternative to mere storage. Incineration, however, has several important drawbacks. First, (poly)halocompounds are fire-resistant or simply do not burn, and to ensure destruction, high temperatures ($\geq 1300°C$) and adequate dwell times (≥ 1 second) must be attained. Burning of chlorinated organic matter inevitably yields corrosive $HCl-H_2O$ mixtures, adding to the materials problem. Altogether, the cost of incineration - on land - is high, and with chlorine contents of $> 30\%$ by weight, easily surpasses $ 500 per tonne. Furthermore, great care has to be taken to ensure safe operation. A special feature of burning chlorine-containing mixtures is the formation and possible emission of hazardous chlorinated dibenzo-p-dioxins and dibenzofurans.[1] At present, there is still insufficient insight into the details of the chemistry behind. It seems logical to suspect (chlorinated) benzenes and/or phenols to be their precursors. Formation of such aromatic compounds is a common phenomenon when heating organic matter. It has, for example, been established that even incineration of polyethylene in the presence of sodium chloride yields a full range of (poly)chlorinated benzenes.[2] Conversion, including chlorination, of relevant products of incomplete combustion (PIC's), can also take place in the final stages of the incineration process, at relatively low temperatures and catalyzed by metal salts present in the fly ash.[3]

There is clearly a need for a better method of destruction of organic chlorine derivatives, which should work also if the chemical composition of the waste is complex, and if the halogen content is high. Biological treatment, though promising in treating polluted water or soil[4], does not seem to qualify, and this is likely to hold for any sophisticated chemical method. Whereas, for example, catalytic hydrogenation can be a good option in special cases, such as mineral oils contaminated with PCB, or when one wishes to convert a well-defined industrial by-product, treatment of 'real' wastes, of varying, complex chemical and/or physical constitution, will be thwarted by catalytic deactivation, or, alternatively, will call for tedious and costly pre-treatments.

Pyrolysis per se, heating in absence of oxygen, which has caught attention in conversion of biomass or domestic refuse, seems unattractive when chlorinated waste is present, due to survival of chlorinated benzene derivatives.[5]

Thermal hydrogenolysis

We have found[6] that thermal hydrogenolysis, heating in a hydrogen atmosphere, is an effective and versatile method for conversion of any type of organochlorine compound including benzenes[7,8], PCB's[9], chloroethylenes[10] and C_1 derivatives from CCl_4 to CH_3Cl.[11] The C-Cl derivatives are converted into HCl according to the general equation (1), X = Cl.

$$R - X + H_2 \xrightarrow[1-10'']{500-950°C} R - H + H - X \qquad (1)$$

Benzene derivatives give high yields of benzene, and non-aromatic compounds lead to CH_4 and C_2 - hydrocarbons as major products. Experiments with a large variety of other benzene derivatives have shown that functionalities such as X = F, CN, NH_2, OH are 'mineralized' according to equation (1) with comparable or higher rates. The reaction mechanism involves attack of C - X bonds by <u>atomic</u> hydrogen (H·), which is present in thermal equilibrium with H_2 molecules. The reaction, therefore, cannot be subject to anti-catalysis, and - at adequate conditions as to temperature and dwell time - will proceed to the degree of completion as governed by thermodynamics. As a typical example, hydrodechlorination of chlorobenzene, at ca. 900°C, with $H_2/Cl=3$ leads to 99,95% HCl. From a tetrachloro derivative, the remaining fraction will be $(0.0005)^4$ or $\ll 1$ ppb, as has been demonstrated experimentally. Particularly pleasing were the results with hexachlorobenzene[8], a most recalcitrant waste compound; with $H_2/Cl \cong 5$, the reaction was remarkably clean even at 975°C, giving HCl, benzene, biphenyl (4%), chlorobenzene (1%) and some PAH.

Coal-mediated reactions

Substrates other than arenes may lead to soot, even when volatile and hence reacting in the homogeneous gas phase. A prominent example of this category is hexachlorobutadiene. In general, using complex mixtures, or real industrial wastes, some sooting seems inevitable. This problem could be solved by conducting the hydrogenolysis reaction over a bed of coal particles.[12] Only volatile product hydrocarbons leave the reactor - together with HCl, etc. - and substrates prone to condensation and sooting are carbonized.

These findings have led us to investigate the possibilities of combining hydrogenolysis of waste with coal gasification as a source of the reducing gas. Indeed, this worked very well on a lab scale, using a two-stage moving bed reactor system. Components such as CO, CO_2 and H_2O (steam) did not interfere with the hydrogenolysis reaction, and no unwanted side products (dioxins, dibenzofurans) were formed.

Furthermore, proper types of carbon were found to be catalytically active - even without using traditional metal (salt) catalysts.[12] This has opened the way to effective, and complete, dechlorination at temperatures well below 600°C.

Further research and development, in co-operation with the Technical University of Delft, The Netherlands, aims at a pilotplant reactor for 'broad spectrum' waste treatment and a full process design for practical application by 1990.

References

1. Rappe, C., Nygren, M., Lindström, G., Buser, H.R., Blaser, O. and Wüthrich, C. (1987). Polychlorinated dibenzofurans and dibenzo-p-dioxins and other chlorinated contaminants in cow milk from various locations in Switzerland, Environ. Sci. Technol. 21, 964-970.

2. Lahaniatis, E.S., Road, R., Bieniek, D., Klein, W. and Korte, F. (1981). Formation of chlorinated organic compounds through the combustion of polyethylene in the presence of sodium chloride, Chemosphere 10, 1321-1326.

3. Hagenmaier, H., Kraft, M., Brunner, H. and Haag, R. (1987). Catalytic effects of fly ash from waste incineration facilities on the formation and decomposition of polychlorinated dibenzo-p-dioxins and polychlorinated dibenzofurans, Environ. Sci. Technol. 21, 1080-1084.

4. Abs.: Pap. 194th ACS-Meeting, New Orleans (1987), no's ENVR 44-49.

5. a. Herrler, J. and Wedde, F. (1986). In Recycling International, Berlin.
   b. Piechura, H. (1987). World Conference on Hazardous Waste, Budapest, paper 129.

6. Eur. Pat. Appl. 85.20136; US 770.392.

7. Manion, J.A., Dijks, J.H.M., Mulder, P. and Louw, R. (1988). Studies in gas-phase thermal hydrogenolysis. Part IV. Chlorobenzene and o-dichlorobenzene, Recl. Trav. Chim. Pays-Bas, in press.

8. Ahonkhai, S.I., Louw, R. and Born, J.G.P. (1988). Thermal hydro-dechlorination of hexachlorobenzene, Chemosphere, submitted for publication.

9. Manion, J.A., Mulder, P. and Louw, R. (1985). Gas-phase hydrogenolysis of polychlorobiphenyls, Environ. Sci. Technol. 19, 280-282.

10. Manion, J.A. and Louw, R. (1988). Gas-phase hydrogenolysis of chloroethylene: Rates, products and computer modeling, J. Chem. Soc. Perkin Trans. II, in press.

11. Spronk, R. and Louw, R.; to be published.

12. Neth. Pat. Appl. 87.01998.

HAZARDOUS SLUDGE MANAGEMENT IN ITALY

G. Mininni, V. Lotito, M. Santori, L. Spinosa
C.N.R. - Istituto di Ricerca sulle Acque, Roma and Bari

1. Legislation

Treatment and disposal of hazardous wastes are regulated in Italy by the decree D.P.R. 915 enacted on September 10, 1982 (in conformity with the EEC Directive No. 75/442, 76/403 and 78/319), by the Technical Resolutions of the Inter-Ministerial Committee enacted on July 27, 1984 and other laws from these derived. According to these regulations, wastes are classified as:

A) Urban Wastes
* Non-bulky wastes from households or other urban activities.
* Bulky wastes such as furniture, electric appliances commonly used in households or other urban activities.
* Any wastes collected from:
  - streets and public areas
  - streets and private areas which are of public use
  - sea beaches, lake beaches, and river shores.

B) Special Wastes
* Any wastes that are not classified as urban solid wastes due to either their quality or quantity.
* Wastes from industrial activities.
* Wastes from agricultural activities.
* Wastes from artisan and commercial activities.
* Wastes from hospitals that are not classified as urban solid wastes.
* Building, demolition and excavation materials; out-of-use machineries and equipment.
* Motor vehicles and their parts that are no longer in use.
* Any residues from waste and wastewater treatments (e.g. sludge).

Any special wastes from industrial, agricultural, artisan, or commercial activities and from hospitals and any residue from waste and wastewater treatments are classified as TOXIC AND NOXIOUS if any of the following conditions are met:
- Any of the elements or compounds, listed in the annexed

Tables of the Technical Resolutions, is present at a concentration greater than the standard one (see table 1).
If they derive from certain activities (see table 2) unless the producer can prove that the waste does not contain any of the listed elements or compounds in a concentration greater than the standard one.

Regional Authorities are charged of issuing excutive plans regarding the solid wastes treatment and disposal. To this end they must identify types and quantities of wastes, decide the plants sites, approve projects for the urban and special wastes treatment plants as well as the companies charged for disposal sevices. Moreover Regional Authorities must establish local Regulations in order to **organize** and control the activities of wastes disposal. **Most of** the Regional Authorities have not issued till now these plans. Therefore it is **difficult to** estimate the production **of speci**al wastes, especially the hazardous ones.

Some sources (Infraconsult, 1985; Misiti, 1987; Nels, 1986; Wilson & Forester, 1987) indicate a yearly amount of hazardous wastes ranging between 2 and 5 Mt out of 35-45 Mt of industrial wastes.

According to OECD about 1/3 of industrial waste is to be considered as hazardous, thus leading to an estimated hazardous waste production until 15 Mt per year in Italy.

Table 1 - Some examples of concentration standards (mg/l)	
Antimony and its compounds	5,000-50,000
Arsenic and its compounds	100
Berillium and its compounds	500
Cadmium and its compounds	100
Chromium (VI) and its compounds	100
Mercury and its compounds	5,000
Lead Alchyls	100
Copper (soluble compounds)	5,000
Selenium and its compounds	100
Tellerium and its compounds	100
Phenol and its compounds	500 -50,000
Cyanide (organic and inorganic)	500 -50,000
Organic Halogen compounds except for	100 -50,000
2,3,7,8 tetrachloridedibenzodioxin	0.001
1,2,3,6,7,8 hexachloridedibenzodioxin etc.	0.001
Chloride solvents	100 -50,000
Organic solvents	500 -50,000

Table 2 - Examples of **activities** from which derive toxic and noxious wastes

a) wastes deriving from production of
   - PCB, PCT, PCF
   - chlorinated hydrocarbons
   - pharmaceutical compounds
b) process sludges deriving from
   - galvanic baths containing hexavalent chromium and cyanide
   - wood treatment by means of creosote and pentachlorophenol
   - hardening of metallic **surfaces** by means of cyanide baths
   - oil products storage
   - removal of fats from metallic surfaces by means of chlorinated solvents
   - **removal of emissions derived from steel production in electric furnaces**
c) residues and bottoms of distillation
d) exhaust solution from galvanic baths
e) exhaust solvents

Currently only a small portion of hazardous wastes is processed in centralized plants (approximately 400.000 t/year).

2. Stabilization/solidification plants

The plant operated in Modena by the local Agency responsible for urban solid wastes collection and treatment is based on the Soliroc process. The plant has a capacity of 30,000 t/y of sludge or 25,000 t/y of concentrated liquids. Prevalently inorganic wastes **are transformed into silicates** then landfilled. The acid liquids are introduced into an "acid phase" reactor where blast furnace slag is added to form monomer silicic acid; this solution is then moved to an "alkaline phase" reactor where polymerization takes place and silicates are formed by adding alkaline liquids or lime. Highly toxic sludges are solubilized possibly using recycled acid wastes before being subjected to the above treatment, while low toxic sludges are submitted to cementation only by adding lime or cement and blast furnace slag. Liquids containing cyanides are treated separately

adding hypochlorite and soda for the removal **of cyanides.**
After treatment the alkaline liquids are stored for recycling purposes.

The plant operated near Turin by Servizi Industriali (a private Company) handles about 200,000 t/y of industrial wastes. Liquid and solid wastes are treated by the Chemfix process, during which a **gelatinous** mass is first formed and then solidifies, thus becoming similar to natural materials like zeolites, feldspats, etc. Liquid wastes containing hydrocarbons are treated separately by centrifugation and three phases are obtained: the oily phase is reused as fuel, the aqueous phase is treated biologically and the sludge by the solidification process. All the sludges produced are landfilled.

The centralized plant in Brescia is operated by Ecoservizi (a private Company) and handles more than 100,000 t/y of hazardous wastes. Sludges having a low metal content are directly disposed of in a controlled landfill, while those having a high metal content are previously treated by the Litosintesi process which makes toxic metals insoluble by the addition of specific reagents, such as cement and sodium silicate.

At a centralized plant treating tannery wastewaters near Vicenza a mobile plant, based on the Petrifix process, having a capacity of 150 t/h of sludges with a high concentration of chromium, was used. This process has the advantage of being compatible with the presence of organic matter. The following steps can be distinguished: neutralization, precipitation (of silicates, borates, arsenates, phosphates), adsorption (of Pb, Zr, V, Mo, Te, U, Pt), solidification and disinfection; chemicals used include fly ash, blast furnace slag and cement. The organic matter does not interfere with the reactions, but it is not fixed into the crystalline structure and, therefore, can be leached afterwards.

Leaching tests on stabilized/solidified materials are required in order to select the proper landfill type. A solution of acetic acid (0,5 M) is generally used for leaching (pH lower than 5). If only inorganic solid wastes are disposed of, a saturated solution of CO_2 is used.

If the test is favourable (heavy metals in the leachate within the limits of the standards fixed by the law for the effluents of industrial wastewaters treatment plants in the surface waters) the stabilized material can be disposed of in a controlled landfill less secured and then less expensive. In the other case the landfill must be lined and the beneath soil must have a layer of at least 2 m with a permeability lower than 10^{-7} cm/s.

3. Incineration plants

Very few plants are now in operation in Italy with a total **capacity of about** 200.000-300.000 t/y, while the demand is over 1.000.000 t/y (Toscanini, 1987).
The plants installed at the Montedipe and Monteco factories in Mantua, Ferrara and Porto Marghera (Venice) are briefly described in the following. Some of these are oversized for internal needs and, therefore, utilized for incinerating wastes on commission.
Mantua incineration plant consists of a 12,000 t/year rotary kiln furnace that can be fed with solids, liquids and sludges. The plant is completed by a static afterburner chamber and a scrubber. High-viscosity semi-liquid wastes are previously fluidized in a water-heated chamber and then fed to the furnace.
The Ferrara plant has a capacity of 1,250 kg/h of solids, 500 kg/h of liquids and 1,500 kg/h of sludges; it consists of the following units:
- rotary kiln furnace for solid wastes
- fluidized bed furnace for sludges
- afterburner chamber in which liquids can also be treated
- tower for flue gas washing by alkaline solution.
The Porto Marghera plant has a total capacity of 85,000 t/y. It consists of two fluidized bed furnaces (210 t/d each) for sludges and two static furnaces for liquid wastes (1500 kg/h each). In this plant the recovery of hydrochloric acid solution (28%) from flue gas is of particular interest.
Technical requirements for incineration regard the afterburning of exhaust gases, which must be carried out, according to the following standards:
* Oxygen concentration in exhaust gas from the post-combustion chamber must be greater or equal to 6% in volume.
* Mean gas velocity at the inlet of the post-combustion chamber must be greater or equal to 10 m/s.

* Residence time in the post-combustion chamber must be greater than or equal to 2.
* $CO_2/(CO_2+CO)$ ratio must exceed 0.999.
* Combustion temperature and oxygen concentration must be continuously monitored.
* Combustion temperature requirements are as below:
  - For urban wastes and special wastes with organic chlorine contents not exceeding 2%, the temperature must exceed 950 °C.
  - For urban wastes and special wastes with organic chlorine contents exceeding 2% the temperature must exceed 1200 °C.

4. References

Infraconsult (1985). Definition and classification of toxic and dangerous wastes EEC Contract N. 84-B-6643-11-004-11-N.

Misiti, A. (1987). Problematiche di piano relative allo smaltimento dei fanghi e dei rifiuti tossici e nocivi (Planning problems for hazardous waste disposal), National Seminar on "Hazardous sludge treatment and disposal", Bari, Jan. 20-21.

Nels, C (1986). Hazardous waste incineration, Course on "Incineration of hazardous and non hazardous waste", Center for Professional Advancement, Amsterdam, Apr. 7-10.

Toscanini, C (1987). Il ruolo dell'industria pubblica e privata nello smaltimento dei rifiuti speciali, tossici e nocivi, 25° RAMEL, Viareggio 28 settembre.

Wilson, D.C. & Forester, W.S. (1987). In International Perspectives on Hazardous Waste Management (Ed. Forester, W.S. & Skinner, J.H.) 153-159. Academic Press, London.

HAZARDOUS WASTES MANAGEMENT IN JAPAN

Tadayuki MORISHITA,
Director General,
Water Supply and Environmental
Sanitation Department,
Japanese Government

I   Hazardous Waste Problems in Japan

In Japan, wastes are categorized as municipal solid waste and industrial waste. The term "hazardous wastes" is not defined under the law, but this term applies to such industrial wastes as cinder, sludge, waste acid, waste alkali, slag and dust, found by extraction test to exceed the prescribed hazard limit.

The following substances are subjected to hazard examination: alkyl mercury, mercury or its compounds, cadmium or its compounds, lead or its compounds, organic phosphorus compounds, hexavalent chromium compounds, arsenic or its compounds, cyanide, and PCBs.

Wastes other than the items mentioned above, and those which do not exceed the preset limit, are not subject to hazardous waste control.

Although dioxin and mercury in municipal waste or waste oil in industrial wastes, for instance, are not defined as hazardous waste, they are called "problem wastes" and are treated as carefully as hazardous wastes.

II Some Experiences related to Hazardous (Problem) Wastes
1. PCB problems

In Japan, the use of PCB was prohibited in 1972.

PCB wastes are classified into following three types: (1) **liquid PCB which was first used as thermal medium and then** recalled by its manufacturers; (2) PCB used as an insulation oil on transformers and others and, (3) PCB used on pressure sensitive paper.

Direction of PCB treatment is as follows:

The demonstration test of liquid PCB by thermal decomposition was conducted by the Environment Agency and favorable results were obtained.

Actually, a private enterprise constructed the incinerator in order to decompose liquid PCB under the following design based on the results of the test by the Environment Agency.

(1)	Treatment capacity	630 kg/h
(2)	Decomposition temperature	1,400 °C
(3)	PCB decomposition rate	99.9999 % or higher
(4)	PCB emission concentration (exhaust gas)	0.01 mg/m
(5)	PCDD and PCDF emission concentration	Quantitative limit or lower

The treatment of liquid PCB is expected to commence this year.

2. Waste Dry Battery

From fall 1983, incineration and reclamation of mercury-containing waste dry batteries have been causing a fear among people as a potential threat to environment. Although research on mercury's environmental concentration and discharge from solid waste treatment processes did not prove the threat, following measures have been taken in order to secure smooth operation of municipal solid waste management.

(1) Mercury batteries (button-shaped) are to be collected and disposed of by battery manufacturers.

(2) Mercury content of **alkaline-manganese** dry batteries is to be reduced to 1/6 of 1983's content by September 1987, by battery manufacturers.

(3) Regional disposal system of collected waste dry batteries has been promoted by the Japan Waste Management Association (JWMA) organized by municipalities with the cooperation of battery manufacturers.

3. Dioxins

In Japan, with the detection from incinerator residue of municipal solid waste management in 1983 fall, dioxins have been known among people in general. Adhoc Council of Experts on Dioxins and Waste Management established the evaluation guideline of dioxins in the field of solid waste treatment (0.1 mg/kg/day as 2, 3, 7, 8-TCDD) and concluded that the maximum potential exposure to incinerator workers and people in general would be lower than the guideline. So far, a variety of investigations and researches have been carried out on dioxin's formation, decomposition and behavior in waste treatment processes, and those show that all data were lower than the evaluation guideline level.

4. Management of Asbestos Wastes

Recently, increasing amounts of waste containing asbestos have been released with the rapid redevelopment of cities and renovation of buildings.

Therefore, the Ministry of Health and Welfare issued temporary guidelines on the management of asbestos waste in order to fill an immediate need, last year, and comprehensive guidelines on the management of construction industrial asbestos waste will be made this year.

III Manufacturers' Self-assessment System on the Disposability of their Products

(1) Problems

With the rapid socio-economic change, many kinds of hardly disposable wastes such as bulky waste refrigerator, waste battery containing mercury and waste spray cans with explosive character have been discharged to municipal solid waste management system, and many municipalities have some difficulties to dispose of these wastes properly.

(2) Guidelines on Self-assessment

The Living Environment Council, an advisory organ to the Minister for Health and Welfare, and the Experts' Committee under the Council made reports which put emphasis on the importance of the manufacturers' self-assessment on the disposability of their products so as not to impair appropriate municipal solid waste disposal.

Based on the reports, the Ministry of Health and Welfare issued the notice on Self-assessment, with the guideline, to enterprisers, local governments and other agencies concerned in December 1987.

INTERACTIONS BETWEEN CLAY MINERALS AND ORGANIC COMPOUNDS AROUND HAZARDOUS WASTE LANDFILLS

W. Smykatz-Kloss and L. Kaeding, Mineralogical Institute of the University, D-7500 Karlsruhe

<u>Abstract.</u> The claystone layer which occurs at the base of an industrial waste landfill has been chemically and mineralogically altered by the influence of sewage waters of the landfill. The primary, not-swelling clay minerals illite and chlorite were transformed to a swelling interstratification of the irregular illite/smectite type. The pollutants originating from the landfill were partly adsorbed by the "Tonstein" clay minerals (e.g. the heavy metals). Partly they migrate through the claystone layer without any hindrance (e.g. the organic compounds). <u>Experimentally</u> it can be shown that swelling clay minerals (e.g. smectites) quickly react with organic compounds of these types which have been found in the studied landfill to form organo-clay complexes. These reactions influence strongly the rheologic behaviour of the clays. Non-swelling clay minerals (e.g. kaolinites, illites) have not been altered during four weeks' experiments.

I Introduction

Recently more and more clays were taken as sealing material for industrial and domestic waste landfills, to prevent the soil and groundwater of the environment from pollution. The reason is the low permeability of clays for water. But up to now little is known about the chemical and physical interactions between clays and sewage waste waters and about the alteration of the clays and their properties by these waters. Therefore the present study concerns a) with field investigations on a natural "Tonstein" (claystone) layer underlying a landfill of hazardous industrial waste (mainly galvanic muds), b) with experimental alterations of clays by means of organic solutions occurring in similar landfills.

II Field studies

The hazardous industrial waste landfill studied is situated in Northern Badenia and shows a natural "Tonstein" layer at its base. This marly claystone layer of Keuper age is some m thick and very low in its porosity and permeability. It should be well suitable as a natural sealing layer to prevent the groundwater and the soil of the environment from being polluted by the sewage waste waters. By comparison with a non-influenced claystone it emerges, that the basic layer has been altered by the waste waters in the following way:

1) The main primary minerals, e.g. the non-swelling species illite and chlorite, have been transformed into an irregular illite/smectite mixed layer (Fig. 1) with swelling properties.

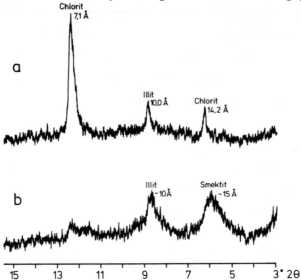

Fig. 1: X-ray diffractograms of an original (a) and an altered claystone from the base of the landfill (b)

This alteration causes a decrease in compaction but an increase in plasticity.

2) The pollutants of the landfill were observed in the underground claystone layer as well. The amounts of heavy metals decrease remarkably with depth (Fig. 2): Ni is only present in the uppermost part of the claystone layer, Pb and Zn migrate down to a depth of 50 cm; Cr and Cu are traceable down to 70 cm. But the organic compounds do not decrease in their amounts with depth (Fig. 2, examples: dichlormethane and trichlorethene).

3) The organic substances are preferably found in the grain size fraction < 2 μm (Fig. 3).

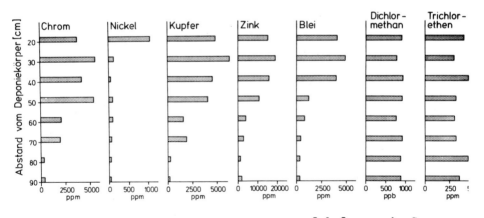

labels are in German

Fig. 2: Contents of heavy metals and organic compounds in the basic layer of the landfill, in dependence on the distance to the lowermost waste; contents in ppm, distance in cm.

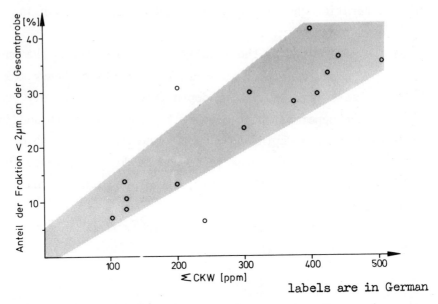

Fig. 3: **Interdependence** between the amounts of organic compounds and the grain size of the basic claystone

III <u>Experimental results</u>

To understand the phenomena observed in the studied and other landfills, e.g. the structural destabilisation of clay minerals and the behaviour of pollutants in the environment of waste disposals, a number of experiments were run. Several clay minerals were contacted with different organic solutions (see Table 1). The results obtained after four weeks were as follows:

1) The non-swelling clay minerals, e.g. kaolinite and illite, do not show any alteration.

2) The main component of <u>bentonites</u>, e.g. the swelling clay mineral species <u>montmorillonite</u>, is strongly influenced by organic solutions (Table 1, Fig. 4). In some cases a <u>disproportionation</u> is observed:

$$\text{organic solution} + \text{montmorillonite} \begin{array}{c} \nearrow \text{illite} \\ \searrow \text{organo-clay complex.} \end{array}$$

Table 1: Basic distances of clay minerals studied after contact with organic compounds (001, in nm)

	kaolinite	illite	smectite
untreated, water-saturated	0,71	1,0	1,52
treated with:			
xylol	0,71	1,0	1,38
toluol	0,71	1,0	1,35
dichlormethane	0,71	1,0	1,70
trichlormethane	0,71	1,0	1,92
tribromemethane	0,71	1,0	1,64
trichlorethylene	0,71	1,0	1,90
tetrachlorethylene	0,71	1,0	1,95

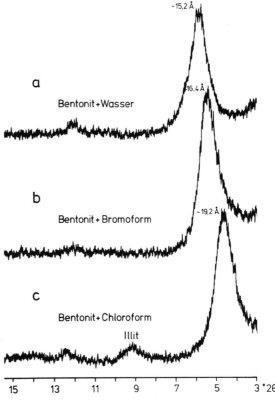

Fig. 4: (001) X-ray spacings of a montmorillonite from a bentonite (in Å) after treatment with water (a), bromoform (b) and chloroform (c)

Labels are in German

In addition to the (001) interference of the organo-smectite the diffuse (001) interference of a badly ordered illite appears (Fig. 4 c).

3) The contact with chloroform results in a larger swelling process compared with the reaction with bromoform.

4) The rheology of organo-clay complexes is remarkably different from that of the clay-water system.

IV Discussion

The most striking alteration effect of clays in the environment of the hazardous waste landfill seems to be the transformation of non-swelling clay minerals (illite, chlorite) into partly swelling irregular interstratifications. This effect is accompanied by an increase of the permeability and of the plasticity as well. The behaviour of the pollutants which leak into the underground clay layer of the landfill is different: the heavy metals decrease with depth, the organic materials do not. It seems that the destabilisation of the primary clay mineral structures (e.g. the formation of the irregular mixed layers) enhance the migration of organic compounds by creating diffusion paths. Smectite species being different due to their crystal chemical composition and to their degree of structural disorder behave differently in contact with organic compounds: The highest reaction rate is observed for Mg-Fe-rich (e.g. trioctahedral) smectites which exhibit a high degree of structural disorder. As a consequence of the present (not yet completely finished) study not only the character of the pollutants of a hazardous waste landfill should be known but the composition and mineralogy of the used sealing clay material as well.

Session 7

Computerized Solid Waste Management

MATHEMATICAL MODEL FOR STRATEGY EVALUATION
OF WASTE MANAGEMENT SYSTEMS

<u>Juha Kaila</u>, KJA Oy, Waste management services
              Helsinki, Finland

1. INTRODUCTION

"Strategy evaluation seeks to combine technologies in such a way as to produce the ´best´ plan against a given criterion, most commonly that of least cost, subject to certain constraints. In operational research, this is an allocation model which **assigns** wastes from sources to sinks and determines the flows. The immediate result is thus a set of waste allocations from each source and to each sink." (Wilson, 1981).

The aim of this paper is to present a mathematical model for strategy evaluation of solid waste management systems.

Figure 1. shows the functional elements which normally could form a solid waste management system. All systems include in practice most or all of the elements on the ´conventional subsystem´ side. The ´resource recovery subsystem´ may in its turn be either a part of the actual waste management system or a parallel but separate resource recovery system.

No model contains all the elements shown in Figure 1. Usually only a few functional elements have been chosen to represent the system to be modelled. In most cases these include transport in collection vehicles and final disposal of waste Quite often some kind of treatment or reloading of waste before final disposal is also included in the models.

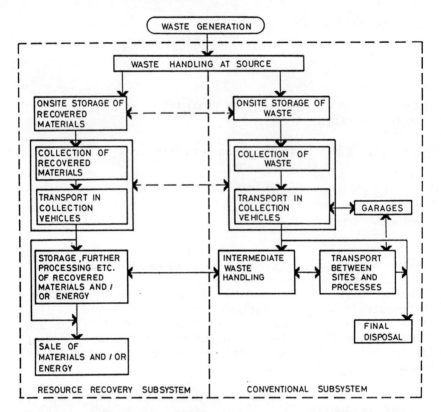

Figure 1. Functional elements of waste management system. (Kaila 1987).

Onsite storage of waste at source has not been included in any strategy evaluation model. Chapman and Berman (1983) have, however, included options for source separation of paper, metal and glass in their model. Although waste transport in collection vehicles is a part of almost every model, the collection phase is normally omitted. It is evident that onsite storage, collection and even transport elements in the system have been either totally ignored or oversimplified, at least compared with the average level of detail in strategy evaluation models. In most models the main interest is focused on waste treatment and more recently also on resource recovery processes and markets for recovered products (e.g. Yale 1978, Chapman and Berman 1983, Wenger and Hansen 1983).

2. MODEL FORMULATION

2.1 Basic assumptions

Each element shown in Figure 1. represents in the model a set of different possible activities and technologies. Some basic assumptions concerning the modelling of these activities and technologies include:
1) waste generation sources can be grouped to form a set of collection areas so that within each area only one type of waste is generated and a collection area may represent a set of sources within a geographical area (e.g. all households within a certain area) or a single source (e.g. wastewater treatment plant producing sludge). Sources generating different types of waste in the same geographical area are thus included in different collection areas.
2) within each functional element the possible activities and technologies are described as a set of alternatives, each with its own performance and cost functions. Thus, for instance, onsite storage is described as a set of alternative onsite storage systems within each collection area so that the resulting system may consist of a combination of these alternatives. For collection and transport elements a set of types of vehicle is defined. Also other means of transport may be included in the set. For each intermediate process, disposal site, and garage a set of possible locations is given and the set of alternative activities and technologies is described for each location.

For solution search the system can be divided into two interconnected subsystems. The first subsystem consists of onsite storage, collection, transport in collection vehicles and garages. The second subsystem consists of the three remaining functional elements, namely intermediate waste handling processes, transport between sites and processes and final disposal.

2.2 Collection and transport system

If it is assumed that the onsite storage alternatives can be grouped so that for each collection area and type of vehicle an average onsite storage arrangement and cost can be defined, then the total unit cost (cost per tonne of waste) can be expressed as:

$$z_{i,b} = a_{i,b} + u_{i,b} + \min_{v \in V_i} \{e_{v,b} + \min_{\lambda \in \Lambda_i} \{w_{i,b,\lambda} + E_{\lambda,i}\}\}, \quad b \in B_i \quad (1)$$

where

$z_{i,b}$ is total unit cost for area i and vehicle of type b
$a_{i,b}$ is unit cost of onsite storage for area i and vehicle of type b
$u_{i,b}$ is unit collection cost with vehicle of type b in area i
$e_{v,b}$ is unit garage cost (per tonne of waste collected with vehicles of type b located in garage v)
$w_{i,b,\lambda}$ is unit transport cost with vehicle of type b from area i to site λ
$E_{\lambda,i}$ is discharge cost per tonne of waste from area i at site λ
B_i^{λ} is the set of possible types of collection vehicle which can be used in area i
V_i is the set of possible garages where such vehicles can be located which can collect wastes from area i
Λ_i is the set of possible discharge sites which can receive waste from area i.

Equation (1) means that for each possible pair of garage and disposal site, the smallest sum of transport and discharge costs is determined first and the unit cost for the garage is then added. The smallest one of these sums is chosen and the collection and onsite storage costs are added. When the total unit costs $z_{i,b}$ have been determined for each area, the amount of waste **to be collected from each area with each type** of vehicle can be determined.

2.3 Processing and disposal system

In order to determine the discharge costs $E_{\lambda,i}$ in equation (1) the least-cost chain of processes from each discharge site to a disposal site must be found for each type of waste. A disposal site itself may be a discharge site and from an intermediate discharge site the number of processes included in possible chains varies. However, in any single chain the number and the order of processes included are fixed.

All intermediate processes and disposal sites can thus be arranged according to their possible positions in alternative **chains so** that all disposal sites are in the last position. All those intermediate processes from which it is possible to transport waste directly to disposal sites are arranged to the preceding position. This procedure is continued until the first position of each alternative chain is reached.

If every process and transport from one process/site to another is regarded as an independent activity, i.e. the costs associated with any activity depend only on the quantities of different types of waste entering that activity, the problem can be solved with dynamic programming techniques as a set of optimal path problems (Bellman and Dreyfus 1962).

If the possible positions for processes are denoted as φ ($\varphi = 1,\ldots,\Phi$), then all disposal sites are at position Φ and cost equal to discharge cost $h_{\Phi,j,\xi}$ is equal to $e_{j,\xi}$ which is the net cost per tonne of waste of type ξ at site j.

For any intermediate process at position φ, where $\varphi < \Phi$ the cost equal to disharge cost at that position can be expressed as:

$$h_{\varphi,m,\xi} = e_{m,\xi} + \sum_{\xi' \in \Xi_m} (r_{m,\xi,\xi'} \min_{m' \in M_{\varphi+1}} \{w_{m,m',\xi'} + h_{\varphi+1,m',\xi'}\}) \;,\; m \in M_\varphi \quad (2)$$

where

$h_{\varphi,m,\xi}$ is cost equal to discharge for waste of type ξ in process m at position φ

$e_{m,\xi}$ is net cost per tonne of waste of type ξ entering process m

$r_{m,\xi,\xi'}$ is reduction factor for waste of type ξ in process m producing waste of type ξ' ($0 < r_{m,\xi,\xi'} \leq 1$)

$w_{m,m',\xi'}$ is transport cost per tonne of waste of type ξ' from process m to process m' (m' is a disposal site when $\varphi + 1 = \Phi$ if transport of waste of type ξ' from m to m' is not allowed, then $w_{m,m',\xi'} = \infty$)

M_φ is set of processes at position φ

Ξ_m is set of types of waste leaving process m.

Starting with position $\varphi = \Phi - 1$ and continuing until $\varphi = 1$ the cost equal to disharge cost at each position for each intermediate process and waste type can thus be determined using equation (2).

Disharge cost ($E_{\lambda,\xi}$) for each intermediate process included in the set of possible disharge sites is:

$$E_{\lambda,\xi} = \min_{\varphi \in \Phi_m} \{h_{\varphi,m,\xi}\} \;,\; m \hat{=} \lambda \in \Lambda_\xi \quad (3)$$

where Φ_m is the set of positions where process m is located.

A final solution for the processing and disposal system is found directly only if the net processing costs ($e_{m,\xi}$), and transport costs ($w_{m,m',\xi'}$) are constant, or amounts of waste which determine these are known. As this is not the case here, i.e. unit costs are not assumed to be constant and the amounts of waste in question are not known in advance, the solution search must be done by iteration.

3. EXPERIENCE OF MODEL USE

3.1 Model use in planning process

The model has been applied several times during actual planning processes. These applications consist of a wide variety of problems of different nature. **Types of problem which have** been dealt with include, for instance:
- determination of which areas should be reserved for solid waste management in land use plans on both regional and municipal level,
- evaluation of future alternatives of waste processing and disposal,
- development of transfer and transport systems when existing landfills are closed and new ones taken into operation,
- determination of suitable schedules for building transfer stations and processing facilities, and
- estimation of costs and benefits of intermunicipal cooperation in a regional solid waste management system.

The model proved to be a feasible tool for strategy evaluation of solid waste management systems because, in all cases, so many questions could be answered that the use of the model **was well motivated. Additionally, the use of the model helped,** both directly and indirectly, the persons who participated in planning to communicate with each other.

3.2 Effects of problem size

As waste flows through the system determine the solution, the size of the problem (P) is here defined as the number of alternative routes of waste flow. As the solution search is done iteratively, the number of iterations required for finding the solution is a clear indicator of how the model responds to changes in the size of the problem.

Figure 2. presents the number of iterations in some computer runs of different applications.

Figure 2. indicates that the number of iterations remains usually quite close to the minimum. What is even more important is that the number of iterations does not increase with the size of the problem. Thus the computer time is almost linearly dependent on the problem size. This makes the model efficient in solving also large problems.

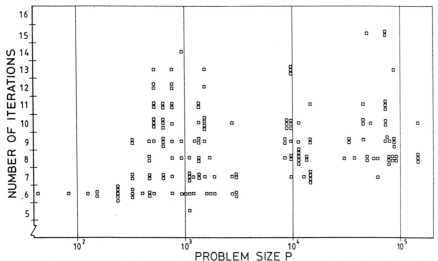

Figure 2 Number of iterations in solution search (Kaila 1987)

4. DISCUSSION

The basic objective in developing the model was to create a tool for practical strategy evaluation applications. Experience of the model use indicate that this goal has been achieved.

The model is more comprehensive than earlier models in taking into account all functional elements of a waste management system. With all elements included it is easier to examine the system as a whole and to recognize the most important interdependencies of the different parts of the system. Additionally, the possibility to include different types of waste into the evaluations makes the model even more realistic.

The possibility to use several types of vehicle simultaneously gives more insight into the transport system because the interdependencies between site locations and the preferred types of vehicle can be demonstrated.

The basic structure of the model is flexible in being independent of the detailed cost and performance functions of functional elements. Many different kinds of function can thus be used, and model details can be modified so that the model is suited for a large variety of problems.

The way in which discharge costs for processes and sites are determined helps to combine the right processes and sites to each other so that cost-efficient chains of intermediate processes and disposal sites are developed. The algorithm does not have to check illogical combinations and alternatives. This is a clear advantage because the relative share of illogical combinations increases with the size of the problem.

The basic approach in model formulation is applicable also outside the field of waste management. The approach would be most suitable for solving problems connected with multistage transport and/or processing system where the flow through the system is analogous to the flow of waste through a waste management system.

REFERENCES

Bellman, R.E., Dreyfus, S.E. 1962. Applied Dynamic Programming. 3.p., Princeton, New Jersey. Princeton University Press. 363 p.

Champman, R.E. Berman, E.B. 1983. The resource recovery planning model: A new tool for solid waste management. National Bureau of Standars. U.S. Dept. of Commerce. NBS Special Publication 657. Washington D.C. 202 p.

Kaila, J. 1987. Mathematical model for strategy evaluation of municipal solid waste management systems. Espoo, Technical Research Centre of Finland. Publication 40. 71 p.

Wenger, R.B., Hansen, J.Aa. 1983. The development of a heat income function for regional solid waste management. Waste Management & Research. Vol. 1, No. 1. pp. 69 - 82.

Wilson, D.C. 1981. Waste Management: Planning, Evaluation, Technologies. Clarendon Press, Oxford University Press. 530 p.

Yale, W.W. 1978. The design of solid waste systems: An application of geometric programming to problems in municipal solid waste management. PhD thesis. Lehigh University. Bethlehem, Pa. 325 p.

COMPUTERIZING IN THE MANAGEMENT OF URBAN CLEANSING MUNICIPALITIES SERVICES AND PLANTS IN ITALY

Dr. A. PERONI   - Managing Director - A.M.I.U. Modena
Eng. A. MAGAGNI - Managing Director - A.M.N.I.U.P. Padova
Dr. P. PREDIERI - Technical Director - A.M.I.U. Bologna

The necessity of knowing how to plan, to manage and to control has become an important requirement of modern society.

The environment, including many changing elements which operate together between them, is a very difficult problem to manage. To control such a process it is necessary to have a lot of information about the conditions, on the trends of the territorial and environmental order.

The information derived must serve for programming the interventions, for a correct knowledge of the laws, and for developing an environmental culture based on reliable information.

We can distinguish 4 elements in the management of information concerning the environment:

- PHYSICS

Geographical, political, territorial, physical objects.

- ANTHROPOLOGICAL-SOCIOLOGICAL

Anthropological and sociological characteristics of the population, of installations and migrations, the statistical data.

- BIOLOGICAL-ENVIRONMENTAL

The natural elements, the flora and fauna, biology, quantity **and quality of chemical dangerous elements, max and min li**mits, reactions between elements, times of treatment and technologies, environmental tolerance.

- LEGISLATIVE

The laws concerning environment: the rules and the limits imposed by laws can be found quickly and can be compared with those of other countries.

The information should help to realize the management of some problems as the following:

- protection of soil, of fauna and flora
- protection of waters and their depuration
- limitation of dangerous chemical substances in the environment
- urban solid wastes, industrial and dangerous wastes: management of services and plants, controls of different phases of treatment
- **treatment, recycling, etc.**
- nuclear wastes and radioactivity
- protection of air pollution
- provisions concerning traffic
- energy exploitment
- protection from traffic noise and from productive activities, etc.

We do not want to list all the instruments suitable for the creation and management of an environmental informative system. We only want to point out that it is necessary to have a lot of instruments if we want to get reliable information: cartography - data bank on population, on laws, on scientific subjects, on chemical products, on environment - control instruments for analysis, models and prototypes - didactic instruments.

The use of computer can be surely useful for environment management, even if it is not the only necessary instrument. We think useful to show briefly the most important fields for the use of informative systems. We will show a plan concerning an informative system for urban cleansing societies.

INFORMATIVE ENVIRONMENTAL SYSTEM PLAN

- BOOK-KEEPING	- MANAGEMENT CONTROL TRADITIONAL SERVICES	- STAFF
- TEXTS MANAGEMENT	- CASH	- SALARIES
- RESOLUTIONS	- CONTRACTS	- PRESENCES CONTROL
- SECRETARY	- SUPPLYING	- CLOTHING AND TURNS MANAGEMENT
- PRIVATE AUTOMATIC	- BUDGET AND CONTROL ON	- IMPROVEMENT OF THE ROUTES

EXCHANGE	BEHALF OF A THIRD PARTY	FOR THE VEHICLES
	- STOREHOUSE WORKSHOP	- CONTAINERS MANAGEMENT
	- PROGRAMMED MAINTENANCE	- VEHICLES MANAGEMENT

PLANTS MANAGEMENTS

- INCINERATORS   - LANDFILLS URBAN, SPECIAL   - DEPURATOR
                    DANGEROUS WASTES
- WEIGHING CONTROL   - COLLECTION STATISTICS   - CHEMICAL ANALYSIS
- WASTES CLASSIFICATION - LANDFILL PLANT CONTROL   - PRODUCTION PROCESS
- REMOTE CONTROL FOR   - PROGRAMMED MAINTENANCE
  PLANTS ON DIFFERENT AREA

WINTER ROAD CONDITIONS
SOIL MANAGEMENT

- URBAN CADASTRE   - REGISTRY OFFICE   - CARTOGRAPHY

ENVIRONMENT MANAGEMENT

- WATERS POLLUTION - SOIL - AIR - ACOUSTIC
- ENVIRONMENTAL SIMULATIONS
- REMOTE CONTROL STATIONS WITH DATA IN REAL TIME

SERVICES MANAGEMENT

- WORK CHARGES
- AUTOMATIC CHECKING FOR QUALITY/QUANTITY OF PRODUCTION ELEMENTS

Management aspects of an informative system: the conditions in Italy

The most attentive operators have already consolidated different methods in the services and plants field, in order to solve rationally the administrative problems.

Here are already some good programs for:
general and directional book-keeping, staff management, the general secretary works, such as the texts, the filing, the documentation instruments and the firm communications management. As far as the technical service is concerned, both for plants and services, only in few societies some procedures have been put in action.

Nevertheless **it is true that during the last years the technical field is particulary attentive for the introduction of** this new management instrument. As far as services are con-

cerned the following computer uses are considered:
- improvement of the routes for the vehicles assigned to urban solid wastes collection;
- the management of all municipal areas (streets, sewers, cesspools, traps, civil settling, position of containers);
- improvement of means usually utilized for the production of services;
- the statistic data processing regarding concentration or spreading of operative units concerning the various services such as collection, sweeping, disinfection, deratization, sewers and street traps cleaning, snow service and so on;
- the automatic weighing of all urban solid wastes containers and immediate knowledge of distances, times and consumptions of vehicles;
- the centralization of staff entry with automation of work turns;
- the programmed maintenance and overhaul of vehicles.

As far as the operating **plants** are concerned the following computer uses are considered:
- the programmed maintenance and overhaul of vehicles;
- control of landfills plants for the treatment of urban solid wastes with terminals installed by Municipalities offices;
- examination of areas for plants location;
- management of productive process;
- management of staff.

The particular computer use in the field of urban services is also examined with regard to the existant structures in the territory considered.

Today there are the suitable conditions to put in action good informative systems in the management of all interventions on the environment.

We think it is useful to talk about the "expert systems" in the field of plants management.

Until now their application has been limited to big societies and in particular fields.

Their spreading and their potentiality lets one hope that they will be adapted also in the municipal societies, above all in the technical fields.

They are programs on computers which store the knowledge and simulate the capacity of decision of human experts: every new event is a source of information for the next resolution.

This instrument will be very useful because it will process quickly a lot of data and connect causes with effects, helping, with this information, to take the best resolution in the shortest possible time.

Wastes classification

The creation of a data bank for wastes is essential for the **pursuing of 3 important aims:**

1 - to plan the whole system of treatment phases
2 - to control the correct respect of laws
3 - to encourage eventual possibilities of recycling.

The wastes classification should contain all the data on which the management of production process and partial or total treatment must be based.

Wastes classification is outlined according to their type and their being dangerous in the following phases:

- production
- **stokages or temporary mass**
- treatment with the authorized process
- transport during the treatment process
- final treatment.

For all the operations in the waste management is possible to have analytical data concerning:

- producers
- conveyers
- treatment societies.

It will be useful to have, considering the different law dispositions concerning the different types of produced wastes (urban, assimilable, dangerous, toxic, noxious, hospital) a classification for the management control of special and to **xic wastes to estimate the produced quantities.**

ENVIRONMENTAL MANAGEMENT

Environmental simulations

It is necessary to make active a series of data bank in order to enrich the management phase of the following services:
- preventive protection
- the examination of environmental tolerance
- the promotion of research and technics for the environment respect
- the development of environmental conscience with the public intervention
- the economic side of environment protection

Some aspects to develop are:
- the nature protection and the landscape protection
- the real and periodical protection of animal and vegetable species
- the environmental impact balance derived from recreative **and tourist structure**
- the control of noxious substances in the soil
- **waters and aquatic habitat protection**
- air protection: air pollution, traffic, heating
- the control of industrial emissions: heating, transformation and energy exploitment
- the struggle against the acoustic pollution: noise in the installations, air traffic, road traffic, railway traffic, industrial activities, yards, recreative activities
- the collection and the waste treatment.

As already pointed out in the urban cleansing field the new instruments are today employed in Italy not only for the traditional administrative services, but also for the management of both services and plants as well.

Session 8

Thermal Treatment of Solid Waste

Section 8

Thermal Treatment of Solid Waste

UPDATE ON GENERATION OF ENERGY FROM MSW
IN THE UNITED STATES

Dr. Ronald J. Alvarez P.E.

The status of the generation of energy from the thermal processing of municipal solid waste (MSW) in the United States as of the spring of 1988 is presented herein together with information on materials recovery. The data contained in this paper is a significant update of that previously presented (1) and demonstrates that increasing quantities of energy are being generated and utilized from the thermal processing of MSW. The material contained herein is limited to full size facilities processing MSW but exclude pilot plant operations and laboratory scale processes.

1. Introduction

Data on operational status, energy generation and usage and materials recovery is presented in Tables 1 & 2 on 206 facilities in the United States from which materials and energy in various forms have been, are being or should be generated in the near future. Table 1 contains data on facilities employing mass burning categorized as waterwall furnace systems, waste heat recovery, or modular incinerators. Data on facilities producing Refuse Derived Fuels (RDF) are given in Table 2.

2. Operational Status

The operational status of the facilities are briefly reviewed in Tables 1 & 2. However, due to the brevity of this paper, only significant changes since the author's last review in 1986 are discussed.

2.1 Mass Burning Facilities With Waterwall Furnace Systems

Of the 83 facilities utilizing waterwall furnace systems, only 31 are shown as either operational or in shakedown and of these, 2 facilities have been closed. The plants in operation represent twice the number operating in 1986.

2.2 Mass Burning Facilities With Waste Heat Recovery

Thirteen of the 21 facilities utilizing waste heat recovery are shown as operational although two are closed representing a 20% increase in operating plants since 1986. Most of the new facilities in this category are in construction or in the contractual stage.

2.3 Mass Burning Facilities With Waste Heat Recovery

Fifty-eight have been identified with modular incinerator

Table 1: Mass Burning Facilities

LOCATION OF FACILITY STATE-CITY/COUNTY(NAME)	YEAR – INITIAL STARTUP	STATUS OPERATIONAL SHAKEDOWN	STATUS CONSTRUCTION	STATUS CONTRACTS	DESIGN CAPACITY tons/day	PROCESS LINES FURNACE/GRATE	WITH SLUDGE	ENERGY USAGE (FORM)
FACILITIES WITH WATERWALL FURNACE SYSTEMS								
AL-HUNTSVILLE	1990*			x	690	2	x	S(s)*
CA-LONG BEACH(SERRF)	1988*			x	1380	3ST		S(e)*
-LOS ANGELES CO.(COMMER)	1987	x			300	1D		S(e)
-STANISLAUS CO.(MODESTO)	1988*			x	800	2M		SI(e)*
CT-BRIDGEPORT	1988		x		2250	3V		SI(e)
-BRISTOL	1987		x		650	2M		SI(e)
-LISBON	1990*			x	500	2R		SI(e)*
-PRESTON(S.E.CONNECT.)	1990*			x	600	2DB		SI(e)*
FL-BAY CO.(PANAMA CITY)	1987	x			510	2OC		S(s&e)
-BROWARD CO.(SOUTHERN)	1989*			x	2250	3V		SI(e)
-HILLSBOROUGH CO.	1987	x			1200	3M		SI(e)
-KEY WEST	1986	x			150	2B		SI(e)
-PINELLAS CO.(PETERSBUR)	1983	x			3000	3M		SI(e)
IL-CHICAGO(NORTHWEST)	1971	x			1600	4M		SI(e)
-FORD HEIGHTS	1991*			x	2500	3M		SI(e)*
IN-INDIANAPOLIS	1988		x		2362	3M	x	SI(s)*
ME-PORTLAND	1988*			x	500	2ST		SI(e)
MD-BALTIMORE(BRESCO)	1985	x			2250	3V		SI(s&e)
MA-BOSTON	1990*			x	1500	2DB		SI(e)*
-BRAINTREE	1971	C			240	2R		S(s)
-HAVERHILL	1989*			x	1650	2V		SI(e)*
-MILLBURY(WORCESTER)	1987	x			1500	2V		S(e)
-NORTH ANDOVER(RESCO)	1985	x			1500	2M		S(e)
-SAUGUS	1975	x			1500	2V		S(e)
MI-JACKSON CO.	1987	x			200	1R		SI(e)
-KENT CO.	1989*				625	2M		SI(s&e)*
MN-HENNEPIN CO.(MINNEIPOL)	1990*			x	1200	2WE		SI(s&e)*
-ROCHESTER CO.(OLMS.CO)	1987	x			200	2R		SI(s&e)
MO-KANSAS CITY	1991*			x	1500	2V		S(e)*
-ST. LOUIS(NORTH CO.)	1990*			x	600	RO		S(s&e)*
NH-CLAREMONT(NH/VT)	1987	x			200	2V		SI(e)
-CONCORD	1989*			x	500	2V		SI(e)*
NJ-BERGEN CO.	1991*			x	3000	4DB		S(e)*
-CAMDEN CO.	1990*			x	1050	3D		S(e)*
-ESSEX CO.(NEWARK)	1990*			x	2277	3DB		SI(s&e)*
-GLOUCESTER CO.	1989*			x	575	2V		SI(e)
-HUDSON CO.	1991*			x	1500	2M		SI(e)*
-PASSAIC CO.	1991*			x	1300	3D		SI(e)*
-PENNSAUKEN	1991*			x	500	2V		SI(e)*
-SOMERSET CO.	1991*			x	700	2M	x	SI(e)*
-UNION CO.	1991*			x	1440	2M		SI(e)*
-WARREN CO.(OXFORD TNSP)	1988*			x	400	2WE		SI(e)*
NY-BABYLON	1988*			x	750	2M		SI(e)*
-DUTCHESS CO.(POUGHKEEP)	1987	x			500	2OC		SI(s&e)
-HEMPSTEAD	1989*			x	2319	3DB		SI(e)*
-HUDSON FALLS	1991*			x	400	2D		SI(e)*
-HUNTINGTON	1989*			x	750	3ET		SI(e)*
-ISLIP	1988*			x	518	2OC		SI(e)*
-LONG BEACH	1989	x			200	1B		SI(e)
-NEW YORK(BROOK.NAVY YD.)	1990*			x	3000	3V		SI(e)*
-NORTH HEMPSTEAD	1991*			x	990	3		SI(e)*
-OCEANSIDE	1965	C			750	3E		I(s&e)
-OYSTER BAY(TOBIDA)	1991*			x	1000	2DB		SI(e)*
-WESTCHESTER	1984	x			2250	3V		SI(e)

Table 1: Mass Burning Facilities (Continued)

LOCATION OF FACILITY STATE-CITY/COUNTY(NAME)	YEAR INITIAL STARTUP	OPER'AL	SHAK DN	CONSTR.	CONTR.	DESIGN CAPACITY tons/day	PROC.LINE FURN/GRA	W/SLUDGE	ENERGY USAGE (FORM)
FACILITIES WITH WATERWALL FURNACE SYSTEMS (Contd.)									
NC-CHARLOTTE(MECKLEN.CO.)	1989*				X	234	2		S(s&e)*
-NEW HANOVER CO.	1984	X				200	2		SI(s&e)
-WAYNE CO.(CELOTEX CO.)	1988*				X	400	2		S(e)*
OK-TULSA	1987	X				1125	3M		SI(s&e)*
OR-MARION CO.(BROOKS)	1986	X				550	2M		SI(e)
PA-BERKS CO.	1991*				X	1200	2M		SI(e)*
-BETHLEHEM(LEHIGH VAL.)	1990*				X	1000	2DB		SI(e)*
-BLUE MT(NORTHAMP.CO.)	1991*				X	1000	2M		SI(e)*
-DELAWARE CO.	1991*				X	2688	60C		SI(s&e)*
-GLENDON BOROUGH	1990*				X	500	2R		SI(e)*
-HARRISBURG	1972	X				720	2M	X	SI(s&e)
-LANCASTER	1990*				X	1200	2M		SI(e)*
-MONTGOMERY CO.(PLYMO)	1991*				X	1200	2ST		SI(e)*
-PHILADELPHIA	1991*				X	2250	3M		SI(s&e)*
-YORK CO.(MANCHEST.TWP)	1990*			X		1344	30C		SI(e)*
PR-SAN JUAN	1991*				X	1050	30C		SI(e)*
RI-JOHNSTON(CENTRAL LAND)	1991*				X	750	2M		SI(e)*
-N.KINGSTON(QUONSET PT)	1990*				X	710	2WE		SI(s&e)*
SC-CHARLESTON	1990*			X		600	2D		S(s&e)*
TN-GALLATIN	1981	X				200	2OC		SI(s&e)
-NASHVILLE	1974	X				1120	4D		SI(s&c)
VA-ALEXANDRIA/ARLINGTON	1987	X				975	3M		S(e)
-HAMPTON(NASA/HAMPTON)	1980	X				200	2D		S(s)
-NORFOLK(U.S. NAVY)	1967	X				360	2D		SI(s)
-PORTSMOUTH(U.S. NAVY)	1976	X				160	2D		S(s)
WA-MARYSVILLE	1991*				X	3000	4M		SI(e)*
-PIERCE CO.	1991*				X	800	2V		SI(e)*
-SNOHOMISH	1991*				X	1000	2R		SI(e)*
-SPOKANE	1991*				X	800	2V		SI(e)*
TOTAL - 83 FACILITIES		28	3	16	36	89912		4	
FACILITIES WITH WASTE HEAT RECOVERY									
CA-NORTH COUNTY(SAN MARC.)	1991*				X	3200	2		S(e)*
-SUSANVILLE(LASSEN C.C)	1985	C				100	1BS		SI(s&e)
FL-BROWARD CO.(NORTHERN)	1990*				X	2200	4VO		S(e)*
-MAYPORT(U.S. NAVY)	1969	X				50	1R		S(s)
-TAMPA(McKAY BAY)	1985	X				1000	4VO		S(s)
GA-SAVANNAH	1987	X				500	2KS		S(s)
MN-PERHAM(QUADRANT)	1987	X				72	2		S(?)
NY-GLEN COVE	1983	X				250	2B	X	SI(e)
-MERRICK	1952	C				600	2B		I(e)
-NEW YORK(BETTS AVE.)	1950	X				1000	4IL		I(s)
-ST.LAWRENCE(OGDENS.)	1990*				X	250	2TR		SI(s&e)*
-SARATOGA	1991*				X	400	2KS		S(e)*
-WASHINGTON CO.	1990*			X		400	2		S(e)*
OH-MONTGOMERY CO.(NORTH)	1988*			X		300	2		S(e)*
TN-LEWISBURG	1980	X				60	1		S(s)
TX-GALVESTON	1991*				X	400	3		SI(e)*
-LUBBOCK	1990*				X	375	3		SI(e)*
UT-DAVIS CO.	1988	X				400	2KS		S(s)&I(e)
VA-GALAX	1985	X				55	1CH		S(s)
-HARRISONBURG	1982	X				100	2B		S(s)
WI-WAUKESHA	1971	X				350	2D		S(s)
TOTAL - 21 FACILITIES		13	0	2	6	12062		1	

Table 1: Mass Burning Facilities (Continued)

LOCATION OF FACILITY STATE-CITY/COUNTY(NAME)	YEAR INITIAL STARTUP	OPER'AL	STATUS SHAK'DN	CONSTR.	CONTR.	DESIGN CAPACITY tons/day	PROC.LIN FURN/GRA	W/SLUDGE	ENERGY USAGE (FORM)
FACILITIES WITH MODULAR INCINERATORS									
AL-HUNTSVILLE(REDSTON)**	1983	X				75	3		S(s)**
-TUSCALOOSA	1984	X				300	4CS		S(s)
AK-SITKA	1985	X				50	2	X	S(s)
AR-BATESVILLE	1981	X				100	4CS		S(s)
-NORTH LITTLE ROCK	1977	X				100	4CS		S(s)
-OSCEOLA	1980	X				50	2CS		N(s)
CA-FREMONT(TRI-CITIES)	1990*			X		480	4VI		S(e)*
-UKIAH	1990*			X	X	100	4		S(e)*
CT-NEW HAVEN	1988*				X	450	5CS		S(e)
-WALLINGTONFORD	1988*			X		420	4VI		S(s&e)*
-WINDHAM	1981	X				108	3CS		S(s&e)
DE-WILMINGTON(PIGEON PT)	1987	X				600	5VI		S(s&e)
FL-JACKSONVILLE(U.S.NAVY)	1981	C				60	3		S(s)
ID-BURLEY(CASSIA CO.)	1981	X				50	2CS		S(s)
KY-CAMPBELLSVILLE	1988*			X		100	4CS		S(s)*
-FT.KNOX(U.S. ARMY)	1986		C			40	1		S(s)
ME-AUBURN	1981	X				200	4CS		S(s&e)
-FRENCHVILLE**	1984	X				50	1		S(s)**
MD-HARTFORD CO.	1988			X		360	4CS		I(s)
MA-PITTSFIELD	1981	X				240	2VI		S(s)
-SPRINGFIELD	1988*			X		360	3VI		S(s&e)*
-WEBSTER	1990*				X	420	4VI	X	N(e)
MI-GENESEE TOWNSHIP	1980	C				100	2		N(s)
MN-COLLEGEVILLE(ST.JOHN)	1981	C	X			60	1		S(s&e)
-DOUGLAS CO.(POPE/DOUG)	1988			X		72	2		S(h)
-FERGUS FALLS	1988			X		95	2JZ		S(s)
-RED WING	1982	X				72	2CS		S(s)
MI-PASCAGOULA	1985	X				150	2		S(s)
MO-FT. LEONARD WOOD	1981	X				75	3		S(s)
MT-LIVINGSTON(PARK CO.)	1982	X				70	2CS		N(s)
NH-DURHAM(LAMPREY)	1980	X				108	3CS		I(s)
-GROVETON	1975	X				24	1		S(s)
-MANCHESTER	1990*				X	560	4VI		S(e)*
-PORTSMOUTH	1982	C				200	4CS		S(h)
NJ-FT. DIX(U.S. ARMY)	1986	X				80	4		S(s)
NY-CATTARAUGUS CO.(CUBA)	1983	X				112	3		S(s)
-ONEIDA CO.	1985	X				200	4		S(s&e)
-OSWEGO CO.	1985	X				200	4CS		S(s&e)
OK-MIAMI	1982	X				108	3CS		N(s)
OR-BROOKINGS	1988*				X	50	2		S(e)*
PA-DELAWARE CO.	1990*				X	50	1		S(e)*
-MONROE CO.(E.STROUDS)	1990*				X	300	3VI		S(e)*
-WESTMORELAND CO.	1987			X		50	2JZ		I(e) &S(s)
SC-HAMPTON	1985	X				270	3CS		S(s)
-JOHNSONVILLE	1981	X				50	2CS		S(s)
TN-DYERSBURG	1979	X				100	2CS		S(s)
TX-CARTHAGE	1985	X				40	1CS		S(s)
-CENTER CITY	1986	X				40	1CS		S(s)
-CLEBURNE	1986	X				115	3		S(s&e)
-GATESVILLE(TX CORR)	1981	X				20	1CS		S(s)
-PALESTINE(TX CORR)	1980	X				7	1CS		I(s)
-SEALY	1984	X				50	1		S(e)
-WAXAHACHIE	1982	X				50	2		S(e)
VT-RUTLAND	1987	X				240	2VI		S(e)
VA-NEWPORT NEWS(FT.EUST)	1980	X				40	2CS		S(s)
-SALEM	1979	X				100	4CS		S(s)
WA-BELLINGHAM	1986	X				100	2CS		S(?)(e)
WI-BARRON CO.	1986	X				80	2CS		S(s&e)
TOTAL - 58 FACILITIES		42	5	2	9	8751		2	

Table 2: Facilities Producing Refuse Derived Fuels (RDF)

LOCATION OF FACILITY STATE-CITY/COUNTY(NAME)	YEAR INITIAL STARTUP	OPER'AL	SHAK'DN	CONSTR.	CONTR.	DESIGN CAPACITY tons/day	PROC.LIN	W/SLUDGE	ENERGY USAGE (FORM)	MATERIAL RECOVERY
CA-SAN DIEGO(NORTH CO.)	1990*				x	1600	1		S(e)*	fga
CT-BRIDGEPORT	1979		C			1800	2		S(e)	fga
-MID-CONN(HARTFORD)	1988			x		2000	2		S(e)	f
DE-DRP(WILMINGTON)	1988			x		600	2	x	S(s&e)	fgan
-NEW CASTLE CO	1972	C				1000	2		S(e)	f
FL-DADE CO.(BANYAN)**	1981	x				1000	1		S(e)**	fganm
-DADE CO.(RESOUR. REC)	1981	x				3000	4		SI(s&e)	f a
-LAKELAND	1983	x				300	1		S(e)	f
-POMPANO BEACH(U.S.DOE)	1978		C			100	1A	x	S(g)	f
-WEST PALM BEACH	1989*					2000	3		S(e)*	f a
HI-HONOLULU	1989*			x		1800	2		S(e)*	f
IL-CHICAGO(S.W.SUPPLEM)**	1978	x				1000	2		S(e)**	f
IA-AMES	1975	x				200	1		S(e)	f
LA-NEW ORLEANS(RECOV I)	1976		C			650	2		N	f
ME-BIDDEFORD	1987	x				600	2		S(e)	f n
-PENOBSCOT(ORRINGTON)	1988			X		600	2		S(e)	f n
MD-BALTIMORE CO.	1976	x				1200	2		S(e)	fg
MA-HAVERHILL & LAWRENCE	1984	x				1300	2		SI(s&e)	f g
-E. BRIDGEWATER	1973		C			500	1		N	f
-ROCHESTER(SEMASS)	1988*			x		1800	3		S(e)*	f n
MI-DETROIT	1988*			x		3300	3		S(s&e)*	f
MN-DULUTH	1985	x				400	1		I(s)*	f a
-ELK RIVER(ANOKA CO.)	1989*			x		1500	2		SI(e)*	f n
-NEWPORT(RAMSEY & WASH)	1987	x				1000	2		SI(e)*	f n
NV-RENO	1990*					1400	4F		S(s&e)*	fga
NY-ALBANY(ANSWERS)	1980	x				750	2		S(s)	f
-HEMPSTEAD(RESOUR.REC)	1978		C			2000	2		SI(e)	fganm
-MONROE CO**	1982			x		2000	2		S(e)**	f
-NIAGARA FALLS(OCCID)	1980	x				2200	3		SI(e)	f
-ROCHESTER(KODAK PARK)	1970	x				180	1		I(s)	f
OH-AKRON	1979	x				1000	2		SI(s&e)	f
-COLUMBUS	1983	x				2000	2		S(e)	f
-GAHANNA	1981	x				1000	2		S(e)	fga
OK-OKLAHOMA CITY**	1978					500	1		S(e)**	f
-LANE CO.	1978		C			500	1		N	f
PA-ERIE	1990*				x	850	2		S(s&e)*	fga
-EASTON(CPM ENER.SYST)	1987	x				300	1		S(s)	f
VA-PETERSBURG	1989*				x	200	1		S(e)*	f a
-PORTSMOUTH(S.E.TIDE.)	1988			x		2000	3		S(s&e)	f a
-RICHMOND(HENRICO CO)**	1984	x				400	2		L**	f a m
WA-TACOMA	1989*				x	500	1		S(e)*	f
WI-LA CROSSE	1988*			x		300	1F		SI(e)*	f n
-MADISON	1978	x				400	1		S(s&e)	f
-MILWAUKEE(AMERICOLOGY)	1977		C			1600	2		S(e)	fga
TOTAL - 44 FACILITIES		26	8	6	4	50030		2		

EXPLANATION OF SYMBOLS IN TABLES 1 & 2

* Estimated
** Not operating now
a Aluminum
c Chilled water
e Electrical
f Ferrous
g Methane gas S(m) or glass, MR
m Mixed metals
n Non-ferrous
1 ton/day(English Units) = 0.90718 t/d (SI Units)

A Anaerobic Digest.
C Closed
F Fluid Bed Incin.
I In-plant usage
L Limited sales
N No sales
S Sold/Contract
B Morse Bolger
BS Bruun & Søren.
CH C&H
CS Consumat Syst.
D Detroit Recip.
DB Deutsch Bab. Air.
E Flynn & Enrich

ET EVT Corp
IL Illinois Rolling
JZ John Zink System
KS Katy-Seghers
M Martin System
OC West.O'Connor Rot
R Riley Energy
RO Rotary Combustor
ST Steinmuller Gmbh.
TR Triga
V Von Roll System
VO Volund System
WE Widmer+Ernst

compared to 44 in 1986. Another ten plants have come on line with 5 more in shakedown which represents an almost 50% increase in operational plants since 1986. However, 3 facilities have subsequently closed and 2 more plants have lost their steam customers.

2.4 Facilities Producing Refuse Derived Fuels

Thirty-four facilities producing RDF are in operational or shakedown compared with 28 in 1986. The new facilities represent what might be termed second generation RDF facilities. Eight of these 34 facilities have been closed and another 4 are not operating, generally due to no customers for the RDF.

3.0 Energy Generation And Usage

Examination of the data in Tables 1 & 2 demonstrates that the energy generated in the form of steam or electricity is being utilized from nearly all operating facilities with the **exception of RDF plants constructed without dedicated boilers** or contracts for sale of RDF.

4.0 Materials Recovery

Very little materials recovery is conducted at mass burning plants with the exception of some source separation or separation of bulky and non-combustible items. However, Table 2 **shows ferrous metals are recovered at all except one RDF facility.** Glass and aluminum are also separated at some plants.

5.0 Pyrolysis

All four of the facilities constructed utilizing pyrolysis processes have been closed and no new projects have been identified since 1982.

6.0 Summary And Conclusions

Sixty-two additional projects have been identified since the end of 1986. The majority of new projects involve mass burning which in general utilizes waterwall systems in the larger urban areas and in modular incinerators in the less densely populated suburban areas representing a trend evident over the past 6 or more years. A number of second generation RDF plants are in shakedown or early stages of operation. Their success or failure over the next several years together with those plants which came on line in the mid 1980's will probably dictate future use of this concept in the United States.

References

Alvarez, R.J. (1986). Status of Energy Generation from Municipal Solid Waste in the United States, Proceedings of 5th International Recycling Congress, 1, 93-98.

PRETREATED REFUSE INCINERATION, IMPACT ON ENERGY RECOVERY, COSTS AND EMISSIONS

J R Barton, D Scott, Warren Spring Laboratory, Stevenage, Herts, UK

1. INTRODUCTION

The general trend in the development of incineration technology to meet improved environmental standards has concentrated on improving abatement equipment. Whilst these developments are needed to meet the more stringent emission limits now being set in many countries, there is also scope for improving combustion characteristics and for reducing emissions by modifying the feed input to incineration by selectively removing certain size fractions or components. This can be achieved by source segregation eg to separate plastics and batteries (to reduce chlorine and heavy metal contents respectively) or by mechanical sorting of the raw input eg screening to remove batteries and non combustible materials. If such processes also enhance the calorific value of the waste, cost savings may be achieved by ensuring better utilisation of the high capital cost of incineration and abatement equipments.

The paper summarises the effect of trommel screening the raw refuse to remove fines (below 40 mm) prior to burning the topsize at the Sheffield incinerator, UK. A detailed report (1) of the trial is available from Warren Spring Laboratory.

2. REFUSE PRETREATMENT AT DONCASTER

Analyses of the refuse normally burnt at the Sheffield incinerator indicated that a significant increase in fuel value would be achieved by simply removing materials below 40 mm in size (see Table 1).

However, to undertake a realistic trial to determine how fines removal would effect combustion, it was necessary to process off-site at the Doncaster RDF plant, where appropriate equipment was available. In fact, for logistic reasons, it also proved necessary to process Doncaster rather than Sheffield refuse and Table 1 shows the separation achieved by using the rotary screen (or trommel) installed at Doncaster.

Table 1. Assays of Sheffield and Doncaster Refuse Samples

Size Range mm	Wt Distbn %	Assay Values (as rec'd)			Distbn (as rec'd)		
		Moisture wt %	Ash wt %	Gross CV MJ kg^{-1}	Moisture %	Ash %	Gross CV %
Sheffield							
+40	70.6	23.0	26.2	11.4	51	75	83
-40	29.4	52.7	20.8	5.6	49	25	17
TOTAL	100.0	31.7	24.6	9.7	100	100	100
Doncaster							
+40	58.9	31.0	21.9	11.0	58	42	70
-40	41.1	32.1	42.8	6.6	42	58	30
TOTAL	100.0	31.4	30.5	9.2	100	100	100

Doncaster waste contained a higher proportion of -40 mm fines than the Sheffield sample and hence a lower weight of topsize was recovered. However the composition of the topsize, in terms of fuel values, was not markedly different from the anticipated values had Sheffield refuse been used.

3. INCINERATION TRIALS AT SHEFFIELD

157 tonnes of topsize refuse was compacted and delivered to the Sheffield Incinerator for the combustion trial. The test was monitored to obtain mass, energy and emission data and the results compared with burning the normal, untreated Sheffield waste.

3.1 Mass and Energy balance

The results in Table 2 show that the mass throughput rating of the grate was maintained at approximately 10 th^{-1}, and the weight loss on combustion increased from 68.2% to 76.6%. Bottom ash flow rates reduced but higher EP ash flow rates and particulate emissions were recorded.

The results in Table 3 show that although the thermal input increased by only 17%, the steam generation rate increased by over 30% from 56,000 MJh^{-1} to 75,000 MJh^{-1} reflecting the increase in overall boiler efficiency from 55.3% to 63.4%. This was mainly due to reduced excess air and higher combustion gas temperatures.

3.2 Flue Gas Composition

Table 4 summarizes the mean flue gas analyses, together with gas phase combustion efficiency and it can be seen that excess air has been cut with reduction in oxygen levels from 11.4% during the raw refuse tests to 9.2% for the screened material. This improvement was achieved by limited optimization of furnace air distribution and was reflected by a 40% reduction in carbon monoxide levels and improved gas phase combustion efficiency.

3.3 Emissions

A summary of stack emissions is given in Table 5. They are expressed as emission factors (mg emitted per kg of refuse burned and normalised for feedrate conditions). The emissions for screened refuse are generally higher than for raw refuse.

Total particle, paper char and carbon emissions for screened refuse were 1670, 171 and 248 mg/kg respectively. These are only slightly higher than for the raw refuse, and not a significant increase.

Metal emissions were similar or slightly higher for the screened refuse in the case of lead, cadmium and copper but again these increases were not significant. Chromium and nickel emission factors were more than twice the values but the difference was only significant in the case of nickel.

Hydrogen chloride emissions were only slightly higher for screened refuse (3780 mg/Kg) but sulphur dioxide emissions were significantly higher (1670 mg/Kg).

The emissions of T(4)CDD isomers were higher for screened refuse (252 ng/Kg) but the increase was not significant. However, T(4)CDF emissions were significantly higher for screened refuse (788 ng/Kg).

Table 2. Incineration Trials, Mass Balance Data

	Pre-screened refuse		Raw refuse	
	teh^{-1}	%	teh^{-1}	%
Feed throughput	10.78	100.0	10.05	100.0
Bottom ash	2.16	20.0	2.95	29.4
EP ash	0.35	3.2	0.235	2.3
Particulate emission	0.019	0.2	0.014	0.1
Wt. reduction on combustion	8.25	76.6	6.85	68.2

Table 3. Incineration Trials, Energy Balance Data

	Pre-screened refuse		Raw Refuse	
	MJh^{-1} x10^3	%	MJh^{-1} x10^3	%
Heat in steam	75.15	63.4	55.88	55.3
Heat in ash	2.51	2.1	4.28	4.2
Heat in stack gas	35.71	31.1	35.68	35.3
Radiation losses	5.21	4.4	5.21	5.2
Total	118.58	100.0	101.05	100.0

Table 4. Stack Flue Gas Concns and Combustion Efficiency

		Mean Gas Concns (Dry)				Combustion
		CO ppm	CO_2 %	O_2 %	Moisture %	Efficiency %
Raw refuse trials	mean	149	8.64	11.38	12.70	99.83
	s.d.	(76)	(1.02)	(1.50)	(1.93)	(0.13)
Pre-screened refuse	mean	84	10.35	9.20	14.94	99.92

( ) Standard deviations

3.4 Reasons for Generally Higher Emissions

Although most of the emission factors were slightly higher for screened refuse, emissions would be generally more favourable when normalised on an energy recovery basis. Possible reasons for higher emissions were:

(i) Use of Doncaster screened waste rather than Sheffield waste (this certainly affected sulphur contents).

Table 5. Summary of Stack Emission Factors (mg/Kg)

Analyte	Means Raw	Means Screened	S.D. (Raw)	Max. (Raw)	Min. (Raw)
Tot. part.	1420	1670	705	2724	848
Paper char	167	171	77	277	78
Pb	19.1	19.1	7.9	33.6	11.1
Cd	0.84	0.89	0.36	1.3	0.3
Hg	1.04	–	0.43	1.6	0.6
Cu	3.4	4.7	1.3	5.4	1.8
Cr	1.5	3.2		3.1	0.6
Ni	2.0	4.3	0.98	3.0	1.0
C	230	248	99	368	142
SO2	1160	1670	305	1674	739
HCl	3556	3780	2311	8628	1978
T(4)CDD	183	252	174	474	18
T(4)CDF	304	788	265	768	70

(ii) EP Performance. The electrostatic precipitator generally performed as well on screened refuse as on raw refuse with total particle capture efficiencies of approximately 95%. Also, the volumes of gas to be cleaned were not significantly higher at equivalent temperatures. However, EP Inlet temperatures were generally higher during screened refuse trials and occasionally led to short-term temperature faults in the first field of the EP.

(iii) Higher Furnace Temperatures. Higher furnace temperatures during screened refuse trials (due to higher heat release and feedrate effects) caused a displacement in the distribution of metals within the incinerator effluent streams. Table 6 shows clearly that the proportion of the metals as well as sulphur and chlorine species reporting to the grate ash is generally lower during screened refuse burning, while the proportion in the arrested fly ash rises. This leads to generally higher EP inlet burdens and even if removal efficiency remains constant, stack emission must rise. Better control of feeding and furnace temperature profiles should reduce this problem.

The higher EP inlet temperatures during the screened refuse trials (mean 298 celsius compared with 268 for unscreened) may contribute to increased dioxin and dibenzofuran emissions. Studies (2) have already shown that dioxins and dibenzofurans can be reformed in EPs under these temperature conditions. The higher dioxin and dibenzofuran emissions were accompanied by a 7-fold increase in the rate of dioxin and T(4)CDF mass flow in the arrested fly ash. These problems should be overcome by operating the EP at lower flue gas temperatures.

Table 6. Distribution of Metals and Other Elements in Effluent Stream

Analyte	Mean distribution of species in effluent stream %					
	Grate ash		Fly ash		Stack emission	
	Raw refuse	Screened refuse	Raw refuse	Screened refuse	Raw refuse	Screened refuse
Lead	71.2	50.6	23.4	45.0	5.4	4.4
Cadmium	46.2	31.7	42.5	61.6	11.3	6.7
Copper	91.6	90.5	7.1	7.6	1.3	1.9
Chromium	87.9	75.3	11.0	21.6	1.1	3.1
Nickel	90.4	81.9	8.7	15.4	0.9	2.7
Sulphur	37.4	11.5	20.2	40.6	42.5	47.9
Chlorine	9.2	2.0	20.7	21.3	70.5	76.7
T(4)CDD	0	0	55.3	78.8	44.7	21.2
T(4)CDF	0	0	22.4	56.4	79.3	43.6

4. CONCLUSIONS

The testwork showed that 30% more energy, and hence revenue, could be obtained from the Sheffield Incinerator by burning pre-screened as opposed to raw refuse. A preliminary economic evaluation, detailed in the test report (1), indicated cost savings of up to 20% could be realised. Stack emission levels increased slightly but were generally within the range expected for raw waste burning. Further trials are underway at the Edmonton electricity generating incinerator where the effect of ferrous metal as well as fines removal will be studied.

REFERENCES

1. Barton, J. R., Poll, A. J. Preparation and incineration of screened refuse; preliminary trial, Sheffield, November 1986. WSL report, LR 592(MR).

2. Vogg, H. et al. Recent findings on the formation and decomposition of PCDD/PCDF in MSW incineration. Special Seminar on Emissions of Trace Organics from MSW Incinerators, January, 1987, Copenhagen, Denmark.

Mercury Analyzer for Continuous Surveillance of the Waste Water of Refuse Incineration Plants

H. Braun, Kernforschungszentrum Karlsruhe GmbH, Karlsruhe,
R. Seitner, Dr. Seitner Meß- und Regeltechnik GmbH, Seefeld,
Federal Republic of Germany

1. Introduction to the Mercury Problem

In the incineration of mercury bearing waste, mercury, unlike other heavy metals, is released completely into the offgas because of its high volatility, causing the well-known problems in offgas purification. Mercury is oxidized quantitatively to mercury(II) and bound in the chloride form by the hydrochloric acid gas present and, in the absence of suitable offgas purification procedures, is emitted through the stack. Wet flue gas scrubbing allows this toxic heavy metal to be transferred from the offgas into the acid flue gas scrubbing water. The wet scrubber represents the sink for most of the mercury [1].

The incineration of municipal solid waste in a medium-size refuse incineration plant annually releases up to 500 kg of mercury with the offgas. In the light of these mercury loads, ecological as well as economic reasons make it imperative that mercury, as a valuable material, should be recovered from the scrubbing water. Check analyses conducted on treatment batches are not very effective. Satisfactory surveillance can be achieved only by on-line measuring techniques.

The mercury content on the input side and, hence, also in the flue gas scrubbing water varies greatly. If the operator knew the instantaneous mercury concentration in the scrubber, he could adjust accordingly the dosage of chemicals in waste water treatment. For the time being, chemicals are added for the maximum concentrations expected to occur [2]. In the treatment of flue gas scrubbing water, the legal waste water limit must be observed. As with other pollutants, continuous surveillance must be demanded also in this case.

2. The Hg-MAT 1 Mercury Analyzer

A mercury analyzer is presented which can be used in continuous monitoring of mercury bearing waste water for mercury, irrespective of its initial form. This detection of mercury succeeds only if mercury is present in its elemental form in the gas phase. In solutions, the mercury compound therefore must first be reduced to its elemental stage, which can then be expelled by air. In the Hg-MAT 1 presented, dissolved mercury chloride is converted continuously and completely into detectable elemental mercury by means of stannous chloride and then passed to an integrated atomic absorption spectrometer.

The mercury analyzer is an advanced version of the mercury analyzer made by Dr. Seitner, Meß- und Regeltechnik GmbH, a proven laboratory unit for mercury trace analysis in the µg range, which has been in use for many years. Design modifications and the addition of a dosage and extraction unit allow the equipment to be run automatically and continuously. The process engineering part was developed in cooperation with, and fabricated under license from, Kernforschungszentrum Karlsruhe GmbH.

Only stannous chloride is required as a reagent for operation. The sample and the reagent are passed by two microflow inducers to a reaction vessel in which the mercury, which has been reduced to metal by the addition of air, is converted into the vapor phase (Fig. 1). The extracted mercury vapor is passed to a heated cuvette where the attenuation of a radiation source with the wavelength of 253.7 nm specific to mercury is measured in a familiar way as a direct indication of the mass concentration of mercury.

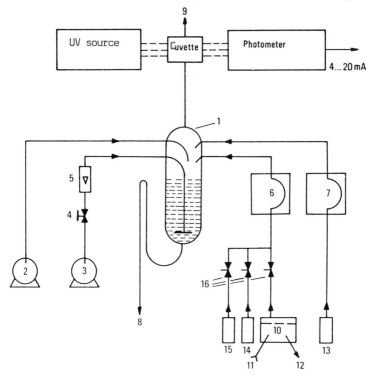

Fig.1: Hg-MAT 1, functional diagram

1 reactor 2 pump for make-up air 3 pump for extraction air 4 needle valve
5 flowmeter 6 pump for sample 7 pump for reagent 8 outlet 9 exhaust air
10 main filter (bypass) 11 sample feed 12 sample discharche 13 reagent
stock 14 blank solution 15 calibration solution 16 solenoid valves

A concentration computer integrated in the unit prepares a digital display of the result in mg/l and outputs it as a process variable in the analog mode. The output has been normalized to 4-20 mA and can be connected to recorders, controllers, or signaling devices. The blank level and the calibration level can be checked by push-button operation to inspect the analyzer. For this purpose, pure water or a calibration solution is fed once instead of the sample. To reduce the delay in indication brought about by long feed lines from the sampling point to the unit, indication is normally carried out in a bypass. The substance to be measured is filtered directly below the analyzer. The Hg-MAT 1 unit, which has been designed specifically for measuring and controlling mercury concentrations in the scrubbing water of the flue gas scrubbers of refuse incineration plants, operates in a measuring range of approx. 0.1-30 mg/l.

3. Treating Flue Gas Scrubbing Water

The mercury analyzer was subjected to a five-day test at the Bamberg refuse heating power plant. Each of the three combustion units of the plant, with a capacity of 7 Mg of refuse/h each, has a self-contained complete flue gas treatment stage consisting of the waste heat boiler, electrostatic filter, and flue gas scrubber. Downstream of each electrostatic filter, a flue gas volume of approx. 35000 std.m^3/h arises. The mercury concentrations amount to approx. 300-600 µg/std.m^3. Flue gas scrubbing removes most of the mercury as well as other substances. A mercury concentration of approx. 3-6 mg/l is established in the scrubber water.

The acid flue gas waste water from all scrubbers is neutralized with milk of lime in a common neutralizing tank. To precipitate mercury, TMT 15 (15% solution of trimercapto-s-triazine) is added at a rate of approx. 6 l/kg of mercury. TMT 15 is fed as a function of the running time of the salt removal pump in a quantity, usually, of 50 ml/m^3 of flue gas waste water. This would allow a maximum of 8 mg Hg/l to be precipitated [2].

The mercury analyzer was used to determine the mercury content directly in the recirculation water of a scrubber. The mercury concentration in the flue gas is known to vary greatly [3]. For the first time, it has been shown that similar fluctuations in the mercury content must be expected to arise also in the scrubber, despite the high capacity of scrubbers (Fig. 2). In addition, the scope of fluctuations was observed to depend on the origins of the household

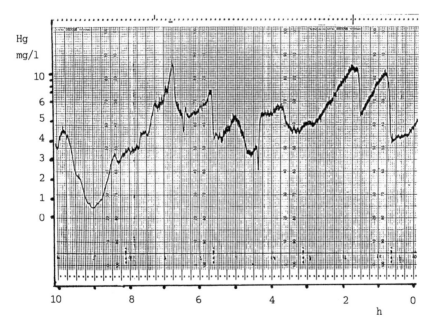

Fig. 2 Mercury concentration in the scrubber water

refuse being incinerated. Now it is possible to explain those cases in which the permissible mercury discharge concentrations had been exceeded: Peak levels are not covered by the uniform dosage of TMT 15. On the other hand, the TM 15 feed may be reduced at lower concentrations.

4. Outlook

The use of the Hg-MAT 1 mercury analyzer in a refuse incineration plant has shown that considerable improvements can be achieved in the continuous treatment of flue gas scrubbing water, if the mercury content is measured continuously. This can be done without great expenditure with the Hg-MAT 1, and it not only has a positive impact on the quality of the waste water, but also produces savings in precipitation chemicals.

5. Literature

[1] Braun, H., Metzger, M. and Vogg, H. (1986). Zur Problematik der Quecksilber-Abscheidung aus Rauchgasen von Müllverbrennungsanlagen, Müll und Abfall 18, 62-71 and 89-95.

[2] Reimann, D.O. (1983). In Müllverbrennung und Rauchgasreinigung (Ed. by K.J. Thoemé-Kozmiensky), 897-914. EF-Verlag für Energie- und Umwelttechnik GmbH, Berlin.

[3] Braun, H., Metzger, M. and Vogg, H. (1987). In Müllverbrennung und Umwelt (Ed. by K.J. Thomé-Kozmiensky, 532-553. EF-Verlag für Energie- und Umwelttechnik GmbH, Berlin.

DETOXIFICATION OF DUSTS DERIVED FROM URBAN SOLID WASTES INCINERATION PLANTS

A. BERBENNI*, C. COLLIVIGNARELLI**, F. NOBILE***, A. BASSETTI****, A. FARNETI****

1. Foreword. The danger of diffusion in the environment of organic and inorganic micropollutants contained in dusts and residual sludges derived from urban solid wastes incineration plants has called for the research of operating methods to eliminate the pollutants and to conceive safe disposal systems. Besides the cinders, the incineration process originates dusts coming from the acids absorption columns, normally neutralized with lime, and from the electrostatic filters. The plants with wet removal systems even generate toxic sludge. Dusts and sludges are usually sent to landfill. Depending on the characteristics of urban solid wastes to be incinerated, dusts, in particular, have high concentration of metals and non-metals mostly under the form of oxides which, due to the size of particles and their wide specific surface, are exposed to the action of external agents and to the leaching action of water. For this reason, before being unloaded in the landfill, dusts must be treated in order to reduce the specific surface which can be attacked and the solubility of compounds. This research is aimed at selecting a treatment system suitable for dusts derived from incineration, by studying the composition of dusts, the possibility to fix their toxic components, the results of the leaching tests and the final destination of the treated products.

2. Physical and chemical characteristics of dusts derived from urban solid wastes incineration plants.
The materials utilized in this research are:
- dusts from the electrostatic filter of an incinerator (1) of urban solid wastes "as they are" with 3,100 kg/h capacity;

---

*   Department of Hydraulic and Environmental Engineering University of Pavia (Italy)
**   Engineering Department - University of Brescia (Italy)
***   Professional   - Milan (Italy)
**** SNAMPROGETTI S.p.A.   - Ecology Division - Fano (PS) (Italy)

dusts from the electrostatic filter and from the acid absorption column, derived from an incinerator (2) with 1500 kg/h capacity which treats the overriddles of urban solid wastes after separation of ferrous metals and organic matter;

Table 1 indicates the physical and chemical characteristics of materials derived from both plants (1) and (2):

The concentrations of metals and non-metals are given in milligrammes per kg of material. All samples contain the following toxic metals: zinc, lead, copper, nickel, cadmium, tin; the following elements are not contained or are below the detection limits: arsenic, mercury, selenium. Therefore, we have carried out a leaching test on the materials in compliance with the Italian laws (test performed using acetic acid).

3. Leaching tests on the materials "as they are".

The results of leaching tests on the materials "as they are" are shown on Fig. 1,2 and 3. From the examination of leachates it comes out that leaching exceeds the allowable limits of the Italian law (Table A of law 319/76) in all samples and for every metal except for tin: therefore, the materials "as they are" have to be classed as toxic and treated before unloading them to the landfill. In particular the dusts from the electrostatic filter present a very high toxicity due to their considerable specific surface. In the dusts from incinerator (2) for example, cadmium and copper exceed 600 times the allowable limits, aluminium 300 times, lead over 1000 times and zinc 400 times.

4. Detoxification of dusts.

The experimental results of leaching tests on the materials "as they are", lead to the conclusion that their dumping can involve a serious danger of water pollution: therefore dusts must absolutely be detoxified. In particular it is well-known that the stabilization-solidification is, from the technical-economic point of view, the most feasible treatment for inorganic refuse, the main risk of which is the release of metals, as in the examined case. Among the various technologies existing at present we have chosen the one which is mostly applied as it is based on the use of common and cost-effective reagents, such as the hydraulic binding agents (cement, silicates).

5. Stabilization-solidification process.

Several stabilization-solidification tests have been performed on dusts coming from the electrostatic filters and from the acid absorption column of the incinerator (2) using various hydraulic binding agents with different applied quantities. A series of test pieces has been prepared using the

following materials:
- waste consisting of the examined dusts
- hydraulic binding agent (cement, liquid silicates)
- addition of water

The use of sludge (with high humidity content) instead of water is being experimented which allows to obtain a satisfying combined disposal.

6. Treated products tests and controls. The tests on treated materials aiming at verifying the fitness of the solidification-stabilization treatment, included the measurement of chemical-physical parameters and the chemical analyses of the leachates.

6.1 Physical-Mechanical Tests. After 28 days seasoning the test pieces were subjected to compression test with free side expansion. The following results were obtained:
- a compressive strength of 105 kg/cm^2 was observed for electrofilter dusts;
- for column residues the strength was 95 kg/cm^2.

The above data are the arithmetical mean of the compression values of the three test pieces for each examined sample.

6.2 Leaching tests on the treated materials. The Italian standards foresee two leaching tests to simulate the behaviour of a waste subject to the leaching action of the meteoric waters; the first considers the chemical attack of the material by acetic acid; the second adopts a solution satured with CO_2. In this case, the first test was adopted, using 0,5 M acetic acid as more severe compared to the test with carbonic acid. It has to be pointed out that the test pieces to be subjected to leaching test have previously been milled to obtain the most severe test conditions; in fact, in this case, the leaching tests could have been performed without grinding the treated materials, as the integrity of their structure (compressive strength) was fully demonstrated during the mechanical tests.

6.2.1 Analysis of toxic metals. Fig. 1, 2 and 3 show the results of the analytical tests on toxic metals performed on the leachates obtained from the materials "as they are" and from materials "detoxified" with hydraulic binder; the metal concentration values can be compared to the respective **Italian** standards (the same standards required for waste waters discharged into surface waters). The analysis of such data shows that the experimented detoxification process applied to the dusts derived from the electrostatic filter and acid absorption column, is extremely effective as regards the stabilization of toxic metals.

6.2.2 Analysis of organic micropollutants (dioxines and furans)
The content of dioxines and furans in the incinerator decreased downstream the flue gas purification system (electrofilter and acid absorption column); therefore, it became necessary to check the possible presence of organic micropollutants in the leachates obtained from treated materials.
Tab. 2 shows the results of this research.
It can be noted that all analytical data regarding dioxines and furans resulted to be below the detection limits.

7. Conclusions. The experimented treatment process regarding residual dusts derived from urban solid wastes incineration plants is based on the action of hydraulic binders and particular additives (Cement, Silicates). The selected treatment process proved to be effective both as regard solidification (as the products presented satisfactory mechanical characteristics) and detoxification (as the leachates obtained after grinding of test pieces were in accordance with the Italian law limits).
The solidified products can be disposed in landfill with less severe construction characteristics compared to those required for disposal of toxic wastes. The process, using different additives according to the composition of the material to be treated, can be also utilized for detoxification of other types of wastes and inorganic toxic sludges.
The research now **aims** at optimizing the process so as to give the treated products water-proofing and mechanical strength characteristics, as to ensure a leachate in accordance with the law limits and the possibility of re-utilizing the products e.g. for road paving and embankment, or for coasts protective works.

Tab. 1 Physical characteristics and chemical composition of samples

Characteristics and parameters	U.M	Plant 1 electrofilter dusts	Plant 2 electrofilter dusts	Plant 2 acid absorption tower dusts
Consistency	–	Very fine dusts	Very fine dusts	Granular
Colour	–	grey	grey	dark-grey
Smell	–	odourless	odourless	odourless
Residue at 105°C	%	99.8	95.9	87.9
Residue at 600°C	%	97.8	84.8	92.7
Calcium	ppm	101000	204000	238000
Sodium	"	48000	41000	16000
Silica	"	136364	35000	69000
Aluminium	"	90300	24000	48000
Zinc	"	7220	19000	7000
Iron	"	55436	15000	35000
Magnesium	"	22800	12000	30000
Lead	"	3139	9600	1800
Copper	"	700	2100	900
Manganese	"	1453	800	1200
Chromium	"	1770	800	1000
Nickel	"	1972	100	500
Cadmium	"	65	400	60
Tin	"	587	300	600
Vanadium	"	150	50	50
Mercury	"	below detection limit		
Arsenic	"	"	"	"
Selenium	"	"	"	"

All chemical compositions are the mean value of two determinations.

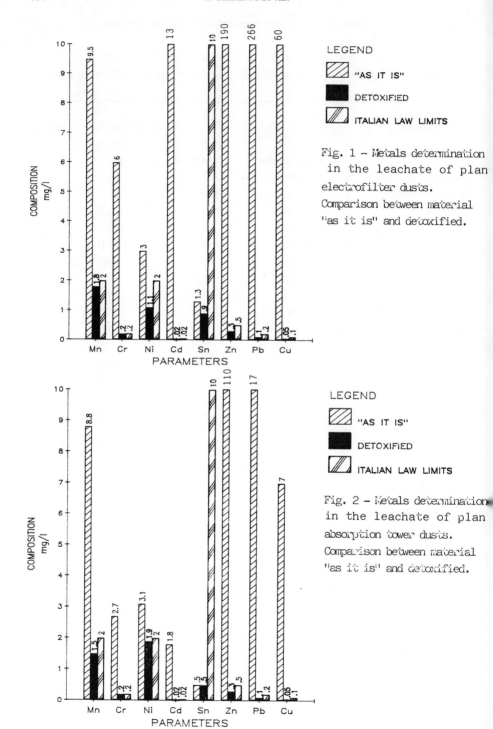

Fig. 1 – Metals determination in the leachate of plan electrofilter dusts. Comparison between material "as it is" and detoxified.

Fig. 2 – Metals determination in the leachate of plan absorption tower dusts. Comparison between material "as it is" and detoxified.

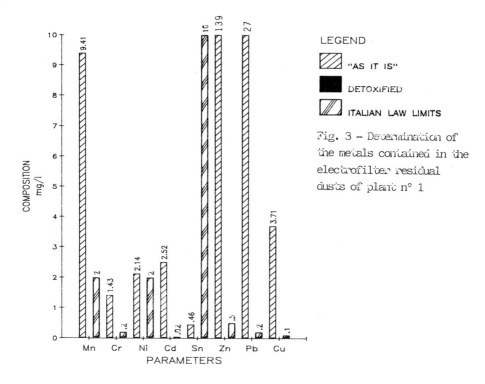

Fig. 3 – Determination of the metals contained in the electrofilter residual dusts of plant n° 1

Tab. 2 Concentration of Polybenzodioxines and polybenzofurans in of detoxified materials (ppt)

	TCDF	PCDF	EsCDF	EpCDF	OCDF
Sample 1	<20	<20	<20	<20	<250
Sample 2	<20	<20	<20	<20	<250

	TCDD	PCDD	EsCDD	EpCDD	OCDD
Sample 1	<20	<20	<20	<20	<250
Sample 2	<20	<20	<20	<20	<250

Sample 1 = Leachate of electrofilter dusts
Sample 2 = Leachate of tower residues

NOTES:
1 – All analytical data are below the detection limit
2 – Analysis performed by the Institute Mario Negri in Milan.

REFERENCES

1. Fanelli R., Farneti A., Carminati A., Benfenati E., Pastorelli R., "Effetti di diverse modalità operative sull'abbattimento degli inquinanti nelle emissioni di un inceneritore di rifiuti solidi urbani" in "Rifiuti speciali tossici nocivi" di A. Frigerio, CSI, Milano (1987).

2. Berbenni P. "Esame del sistema di smaltimento di fanghi tossici provenienti dagli impianti di trattamento di effluenti galvanici e di smalterie." Ed. Quadrio Curzio S.p.A. Milano 1979.

3. Collivignarelli C., Urbini G. "Trattamenti di inertizzazione di reflui speciali, prima dello smaltimento sul terreno". Ingegneria ambientale, $\underline{16}$, 1985 Novembre- Dicembre.

4. Quadrio Curzio P., Nobile F., Berbenni P. "Influence d'une substance ayant une action accélératrice dans le béton projeté armé. Colloque international sur les adjuvants des mortiers et béton. Bruxelles, 30 aout-1 septembre 1967. Rapport IV/13, pag. 213-233 (1967)

5. Berbenni P., "Studio sui processi di innocuizzazione dei fanghi ex - Montedison", AEM, Milano 1982.

6. Norgia R., Nobile F., "Trattamento con leganti idraulici di fanghi contenenti metalli pesanti. Atti Convegno trattamento e smaltimento delle acque residue e dei fanghi industriali" Milano Ed. 1981.

7. IRSA-CNR, "Metodi analitici per i fanghi, Vol. 3, Parametri chimico-fisici", Quad. Ist. Ric. Acque, $\underline{84}$ (1985).

8. IRSA-CNR, "Metodi analitici per le acque", 1972 e aggiornamenti successivi.

9. Wood, Tehobanaglous, "Trace elements in biological waste treatment", J.W.P.C.F., $\underline{47}$, 1933-45, (1975).

10. Berbenni P., Bocconi G., "Die verkannte Grundwesservereiningnung", FEG Informationsblatt 18, (1972).

11   Forstner U., Wittmann G.T.W., "Metal pollution in the aquatic environment", SpringVerlag, Berlin (1979).

12   Berbenni P., "I tests di cessione dei metalli tossici", Atti Convegno IRSA, Bari 18 gennaio 1987.

13   Berbenni P., Nobile F., Bassetti A., Farneti A., "Trattamenti di inertizzazione di polveri e fanghi provenienti da impianti di incenerimento" - CEMPA-SCI-AIH - Convegno Nazionale: "I fanghi e il loro impatto sull'ambiente" - Giardini Naxos 14 - 18/3/88.

THE NEW SAINT OUEN'S INCINERATION PLANT (FRANCE)

Authors : M. Christo DIMITROV - SYCTOM PARIS
M. Daniel BALTZINGER - TIRU SA PARIS

1 - OPERATION'S MAIN FRAME

PARIS and more than sixty municipalities from the close suburbs are gathered into the SYCTOM (Central Association for the Household Refuse Treatment of the Paris Metropolitan Area). Refuse from 4.5 millions people producing 2.1 millions tonnes per year are treated by the SYCTOM.
Municipal wastes are mostly treated in three incineration plants with heat recovery located at SAINT OUEN, ISSY LES MOULINEAUX, IVRY.
TIRU SA Company has been in charge of the operation of these plants since the beginning of 1986 instead of EDF-TIRU the first operator.
Every year, the incineration of 1.8 millions tonnes of waste produces 3 millions tonnes of steam for the Urban Heating and more than 100 000 MWh of electricity.
The age and condition of the first SAINT OUEN plant, built in 1954, has induced the SYCTOM to decide the construction of a new 550 000 tonnes per year capacity plant.
TIRU SA Company was awarded the Engineering contract for this project and will be the Operator when the plant is commissioned.

2 - MAIN CHARACTERISTICS

This new plant's main characteristics are following :

Delivery : a 15 800 m3 storage pit providing a 3 to 5 day plant autonomy.

Incineration : three STEIN INDUSTRIE furnaces, type SITY 2000 each with a per unit capacity of 28 tonnes per hour of waste having a lower calorific value of 8 500 kJ/kg.

Steam production : three STEIN INDUSTRIE energy recovery boilers, producing 40 bars, 380°C superheated steam at the rate of 75 tonnes per hour each.

Electricity generation : a FRAMATOME back-pressure turbogenerator of 13,5 MVA capacity.

Air pollution control : electrostatic precipitators and wet gas scrubbers (DEUTSCHE BABCOCK ANLAGEN process) constructed by TNEE.

3 - MAIN DESIGN CRITERIA

Visual design : great emphasis was put on architectural styling. After competitive bids, the design proposed by the architect S'PACE-MAZAUD was selected.

Plant design : the plant design is fully-enclosed. The different parts of the process including the air pollution control devices are in the main building. The general installation was designed for easy maintenance.

Operating flexibility : in order to optimise the incineration conditions, the process is designed to allow continuous operation down to 40 % of the nominal capacity without deteriorating steam conditions.

Most of the auxiliary equipments has redundancy to ensure maximum availability.

Furnace feeding overhead bridge cranes : the particular design of this equipment allows a feeding cycle of 80 seconds (45 cycles per hour) with a grab capacity of 6 m3. Each of the three movements is controlled by variable speed motors and the whole is coordinated by programmable controller.

Furnace design : the unique design of the SITY 2000 furnace is the result of a collaboration between STEIN INDUSTRIE and TIRU SA.

Boiler design : a minimum fouling and corrosion of the boiler is obtained with design criteria resulting from experience and from the interpretation of the tests conducted in operating plants (ISSY LES MOULINEAUX, IVRY SUR SEINE) The energy recovery boiler design will be 82 % efficient.

Air pollution control devices : the performance of the selected exhaust gas treatment, compared to the French and projected European regulations is as follows :

	Saint Ouen Guaranteed Performance	French Regulation	Projected European Regulation
Particles mg/Nm3	20	50	50
HCl mg/Nm3	20	100	50
SO_x mg/Nm3	140	–	300
Heavy metals mg/Nm3	1	5	5

Each of the three furnace boiler units is equipped with two half gas treatment lines.

Process control : the plant control is centralised, the overhead bridge cranes control is integrated into the main control room.
The plant is controlled by programmable controllers type CGEE ALSTHOM Alspa ZS.

4 - BUILDING PROGRAM

Foundation works started in October 1986 and boiler structure is being erected since February 15, 1988. The first furnace boiler unit is scheduled to begin operation in September 1989.

5 - CONCLUSION

The commissioning in 1989 of the new incineration plant in SAINT OUEN will be an important milestone in the achievement of the aims which SYCTOM has established. The plant will treat more than one quarter of the tonnage produced in PARIS and its suburbs under environmental and return on investiment conditions which constitute a substantial progress not only with respect to the existing SAINT OUEN plant but also when compared to SYCTOM'S two most recent plants. Even though its nominal capacity is less impressive than that of its sister plant at IVRY, the new plant at SAINT OUEN will undoubtedly be in the years to come one of the most interesting industrial projects of its kind in the world.

AN ISOKINETIC SAMPLING SYSTEM FOR CONTINUOUS LONG-TERM
MONITORING OF HAZARDOUS FLUE GAS EMISSIONS

P. Kahanek[1], A. Merz[1], K. Schuy[+2]
[1] Kernforschungszentrum Karlsruhe GmbH (KFK/LIT), FRG
[2] Antechnika Analytische Geräte GmbH, Ettlingen, FRG

1. Introduction. In order to assess the separation efficiency of flue gas cleaning/filtering procedures and to continuously monitor the emission of particle-bound toxics from incineration processes, sampling under strict isokinetic conditions is mandatory. This requirement is particularly relevant when considering solid waste incineration because, as a result of the continuously changing quality of the waste, the quantity and composition of the flue gas may be subject to considerable fluctuations. This applies even more to communal waste incineration facilities.
Within the scope of a research programme "Communal Waste Disposal", inaugurated at the Kernforschungszentrum Karlsruhe, an isokinetic long-term sampling system has been developed. Within a technology transfer scheme with an industrial partner (Antechnika Analytische Geräte GmbH), this system is being prepared for use in large-scale incineration plants.
The system for simultaneous long-term sampling has been installed in the flue gas purification section of the pilot incinerator TAMARA (1) operated at the KFK.

2. The concept of the system. The principle is the development of a long-term sampling system for flue gases, the quality of which in respect of the transported toxics can be evaluated and monitored at any time by means of the sample material which is provided and which can be stored (2.) Although the

system is oriented towards the specific conditions of waste
incineration, it is possible to imagine that it could be used
for isokinetic sampling in many types of multiple-phase
stream. By using microprocessors, it has been possible to
automate, record and process those values which are necessary
for controlling the process. The objective was to considerably simplify the complicated sampling technology.

Figure 1 is a diagram of the modular sampling system. The
partial gas stream is extracted by means of a differential
pressure probe. For removing the particles, the gas stream
can be passed through a cyclone, a plane filter stage or
through a combination of the two. If required, a module can be
connected in the bypass to measure the humidity as a result of
condensation and to absorb gaseous toxics such as mercury,
HCl or SO_2. The complete set of instruments shown in this
figure is recommended for long-term sampling of untreated gas
in order to obtain all analytical information from the
emitted flue gas stream which is important for evaluating
a waste incineration process (3).

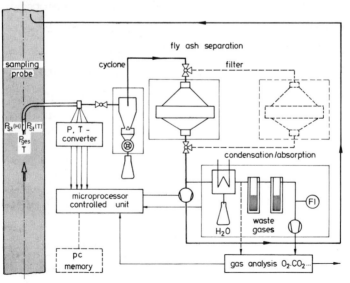

Fig.1: Schematic of the isokinetic sampling system

All components up to the filter outlet, which come into contact with the flue gas, are made of titanium. The probe shaft, the cyclone and filter modules can be heated. The isokinetic regulation is performed by frequency control of the pump motor (max. 400 Hz) with direct drive to a powerful multi-stage suction pump. The underpressure which is attainable is approx. 30 kPa at an exhaust gas throughput of 60 m^3 per hour (density 0.75 kg/m^3).

The physical principle of the differential pressure probe was first suggested by Lajos, Preszler and Marschall (4). All parameters in the gas stream which are relevant for the control process are measured directly at the probe head. These parameters include overall pressure P_{ges}, static pressure in the main stream $P_{St}(H)$ and the partial stream $P_{St}(T)$ as well as the flue gas temperature T. The main and partial stream velocities can be calculated from the dynamic pressures, in a manner analogous to that of the Prandtl Pitot tube. Because of the special geometry, the resistance values dependent on the Reynolds number must be taken into account (5). Calibration measurements in a wind tunnel are necessary in this respect. We have to differentiate between two resistance values:
k for main stream velocity $v_H = k(Re)\sqrt{\frac{2}{3}(P_{ges}-P_{St}(H))}$ and
K for partial stream velocity $v_T = K(Re)\sqrt{\frac{2}{3}(P_{ges}-P_{St}(T))}$
The Reynolds number Re is related to the effective diameter of the probe.

Equality of velocity is present when the partial stream velocity corrected by the probe geometry is equal to the main stream velocity.

Figure 2 shows the components of the long-term sampling system. The sampling probe can be fitted with a heated shaft. In order to achieve optimum coarse dust removal in the cyclone, it is possible to combine various sizes of sampling probes and cyclones. The problem of filter changing has been solved, only short down-times (approx. 3 min.), within a long-term sampling phase of e.g. 6 to 24 hours. The main and partial stream velocities are calculated with consideration to P, T and flue gas humidity and their mathematical relationship with density, Re-number and k(K)-functions at a scanning frequency of 20 Hz. If regulation is necessary, the frequency of the suction pump drive is adjusted by means of digital PID regulator. By the continuous integration of main stream velocity and extracted partial volume stream, the average flue gas volume stream for a sampling phase as well as the gas volume extracted during this time can be directly calculated. The physical design of the extraction probe requires no further flowmeters for determining the volume of the gas stream. The extracted gas volume can be indicated either dry or moist in the operating state or in the standard state. The necessity of automated isokinetic sampling technology becomes clear in figure 3 as a result of measuring values recorded for the dynamic pressures in the main stream $(P_{dyn}(H) = P_{ges} - P_{St}(H))$ and in the extracted partial stream $(P_{dyn}(T) = P_{ges} - P_{St}(T))$. The recordings are taken from sampling

Fig. 2 Sampling System Components

experiments in the flue gas duct of the pilot waste incineration plant TAMARA. According to these recordings, the flue gas volume stream fluctuates both short-term with a periodicity of 1 to 2 minutes as well as over periods of several hours. The partial stream was extracted in accordance with the time function indicated for the dynamic pressure $P_{dyn}(T)$, whereby a delay of 0.30 min. between main stream and partial stream as a result of the recording instrument has to be taken into account.

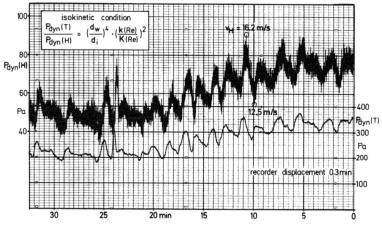

Fig.3 Dynamic pressure P_{dyn} of the flue gas main stream (H) and the sample stream (T)

3. <u>Industrial tests in a waste incineration plant.</u> Subsequent to comprehensive measurements in KFK's own pilot plant TAMARA and in a fluidised solids incineration process for sewage sludge, the sampling system was installed in the flue gas purification system of an industrial waste incineration plant. The objective of these measures was the industrial testing of the system at strategically important positions (point 4: emission; point 1: raw gas i.e. to investigate dioxin formations). The samples taken from positions 2 and 3 can be important if balancing of individual flue gas purification stages and an evaluation of separation efficiency are of paramount importance. The following table summarises typical sampling results.

sampling		flue gas			dust			
position	time [min]	temperature [°C]	sampling volume [Nm^3]	moisture [g/Nm^3]	flue gas concentration [g/Nm^3]	cyclone separation [g]	cyclone efficiency [%]	filter separation [g]
1	44	230	10.715	175.	7.017	69.73	92.7	5.46
2	63	230	10.671	270.	4.671	43.04	86.4	6.80
3	1444	200	215.419	157.	0.021	0.80	17.9	3.67
4	205	115	88.940	198.	0.008	—	—	0.71

Location of sampling positions:
1 raw gas duct between boiler and cyclone prefilter.
2 flue gas duct between spray absorber and electro filter.
3 flue gas duct between electro filter and flue gas scrubber.
4 exhaust duct (emission) between suction blower and stack.

The results of the "industrial trials" should not be considered as long term evaluation figures, rather to demonstrate the universal capability of the sampling system under industrial conditions.

4. Summary. The long-term sampling system works with a special differential pressure probe which ensures that the flue gas partial stream is extracted at a velocity equal to that of the main stream. In this way, sampling errors such as those which were proved by Röthele (6) for simple probes are mainly avoided. Many measuring applications for a wide range of uses are possible as a result of automatic equal velocity extraction and the modular structure of the system components.

References
(1) Merz A., Vogg H., Int.Conf. on Incineration of Hazardous, Radioactive and Mixed Wastes 1988 San Francisco, U.S.A.
(2) Kahanek,P., Merz,A., Chem.-Ing.-Tech.59(1987),12,S.944/946
(3) Merz,A., Tagung "Müllverbrennung und Umwelt" Berlin, 1987
(4) Lajos,T., et.al., Staub-Reinh.Luft 39(1979) 8, S.279/285
(5) Preszler, L., Staub-Reinh.Luft 43 (1983) 8, S.332/336
(6) Röthele, S., Staub-Reinhalt.Luft 42 (1982), Nr.1, S.6/10

Mercury Removal from Waste Gases

K. Keldenich, H. Ruppert, R. Jockers

Bergbau-Forschung GmbH, D 4300 Essen, FRG

In nature mercury can be found in different chemical compounds. The human body takes up elemental mercury mainly from the air. In view of this hazard mercury has to be removed from product- and off-gases of different branches of industry. Some characteristics of the element mercury are listed below:

chemical symbol		Hg
valencies		1;2
rel. mol. mass		200·59
density (293K)	[g/cm^3]	13·59
boiling point	[K]	629·73
melting point	[K]	234·28
enthalpy of melting	[kJ/mol]	2·33
enthalpy of vap. (478K)	[kJ/mol]	59·42
spec. heat capacity (293K)	[J/mol]	27·94
atomic radius	[nm]	0·30
vapour pressure	293K [mm Hg] $1·201*10^{-3}$	13·2 mg/m^3
	313K [mm Hg] $6·079*10^{-3}$	62·5 mg/m^3
	353K [mm Hg] $8·88*10^{-2}$	809·2 mg/m^3
max. permissible conc. at place of work (in FRG)		0· mg/m^3

Mercury is emitted in:
- refuse incinerator plants
- metal smelter plants
- combustion of fossil fuels
- plants for the chloro-alkali-electrolysis

2. Adsorptive Removal of Mercury from Waste Gases

Activated carbons have been used for the adsorptive removal of mercury since a number of years.
In 1934 Stock /1/ successfully used iodine impregnated charcoal to remove mercury by an adsorptive process. These impregnated charcoals were then used by Pütter and Hirsch /2/ as filter material for gas masks. Matsuma /3/ reported that an increase of the iodide content from 5 to 20 % (by weight) increased the adsorption capacity for mercury from 1·4 to 4·8 % (by weight). Impregnation of the activated carbon with sulphuric acid raised the adsorption capacity, too /4/. This effect was considerably strengthened by an additional impregnation with iodide ions /5/. The usefulness of the sulphur impregnation was described by Sinha and Walter /6/ in 1972.

Tests which were carried out at Bergbau-Forschung showed a further distinct improvement of adsorption capacity by impregnation with sulphur and potassium iodide.

2. Tests and Analysis

Under dynamic conditions the adsorbents were investigated using a model gas with $2 \cdot 2 \pm 0 \cdot 4$ mg Hg/m^3 in five parallel adsorbers (Fig. 1). Using AAS the concentration before and after a defined adsorbent layer was measured.
To determine the mercury content the Amalgam Technique was used. Here a defined volume of gas passes a gold/platinum gauze which is inside a quartz tube. This tube is subsequently deamalgamated by heat and then analysis by AAS is carried out.

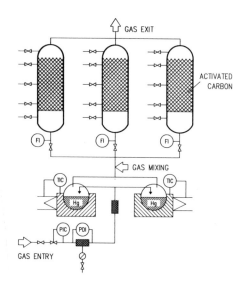

Figure 1: Apparatus

3. Results and Discussion

3.1 Influence of Impregnation

Activated carbon of type D47/4 (BWV) was impregnated by different means. (Table. 2)

Adsorbents	Impregnation
D 47/4	none
D 47/4 KI	2 % potassium iodide
D 47/4 S	11·3 % sulphur
D 47/4 S-KI	11·3 % sulphur + 2 % potassium iodide

Table 2: Tested Adsorbents

These substances in the pore system of the activated carbon increase the purification efficiencies distinctly. Figure 2 shows that the activated carbon impregnated with sulphur and potassium iodide clearly has a better purification efficiency **than all other activated carbons used. After a standing-time** of 4000 h still 99 % of mercury were adsorbed.

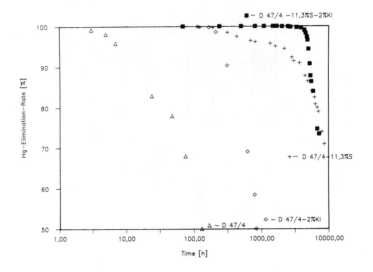

Figure 2: Purification Efficiencies of the Differently Impregnated Activated Carbons

3.2 Influence of Grain Size

The following transport steps are used to describe the kinetics of mercury adsorption:
- Diffusion of the mercury through the layer of lower concentration surrounding the adsorbent grain.
- Diffusion into the pore system of the grain.
- Adsorption of the mercury.

The influence of the different steps on the kinetics was tested on the technical scale by using activated carbons with an identical pore structure but different grain sizes. The structural data are shown in table 3.

As can be seen in figure 3 the activated carbon D45/2-12 % S shows the greatest ability for mercury adsorption. This may be attributed to the grain size of the activated carbon. It has been accepted that the diffusion of the mercury into the layer of lower concentration is the rate determining factor in the adsorption of mercury.

	D 45/2-12% S	D 47/4-11·3% S	D 43/5-11·3% S
Grain Diameter[1] [mm]	2·25	3·89	4·53
Grain Length [mm]	3·17	4·99	4·56
Sulphur Content [%]	12·0	11·3	11·3
He-Density [g/cm^3]	1·971	1·928	2·025
Hg-Density [g/cm^3]	0·807	0·968	0·724
Pore Volume[2] [cm^3/100g]	33·0	27·2	37·2
Pore Volume[3] [cm^3/100g]	40·4	26·6	45·7

[1] average of 20 measurements
[2] determined by benzene adsorption (micropore volume)
[3] determined by Hg-prosimetry (macropore volume)

Table 3: Structural Data of the Activated Carbons

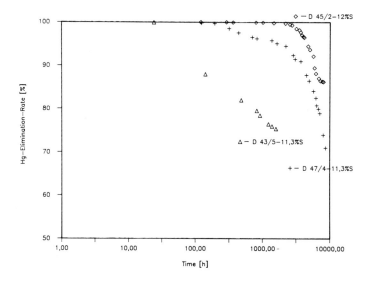

Figure 3: Influence of Grain Size on Purification Efficiency

3.3 Influence of the Gas Velocity

The influence of the contact-time on the adsorption behaviour was investigated by varying the gas velocity. Table 4 shows the variation of gas velocity and its influence on contact-time and Hg-elimination rate in the case of D 47/4 + 11·3 % S.

Gas Velocity [m/s]	Contact-time [s]	Hg-Elimination Rate [bed depth 0·2 m]
0·1	2	96·8
0·2	1	95·9
0·3	0·66	92·5
0·4	0·5	89·6
0·5	0·4	88·0

Table 4: Adsorbent D 47/4 + 11·3 % S
        Standing-Time: 2350 h
        Bed Depth: 0·2 m
        Mercury Concentration: $2·2 \pm 0·4$ mg/m^3

The elimination rate is reduced from 96·8 to 88 % by increasing the gas velocity from 0·1 to 0·5 m/s.

3.4 Influence of Temperature

The effect of the gas temperature on the adsorption equilibrium was investigated in the range of 297 - 373 K. The adsorption behaviour of activated carbon D 47/4 + 11·3 % S at 373 K is very similar to the same activated carbon at 297 K if it has an additional impregnation of potassium iodide (table 5).

Temperature [K]	Hg-Elimination Rate [%]
297	92·8
323	95·8
373	99·9
297	99·9

Table 5: Adsorbent: D 47/4 + 11·3 % S
　　　　　Adsorbent: D 47/4 + 11·3 % S + 2 % KJ
　　　　　Standing-Time:　　1850 h
　　　　　Bed Depth:　　　　0·2 m
　　　　　Mercury Concentration: $2·2 \pm 0·5$ mg/m^3

3.5 Influence of Steam on the Mercury Elimination Rate

Figure 4 shows the influence of steam on the adsorption efficiency at 373 K. Given a bed depth of 0·2 m and a standing-time of 200 h, the elimination rate is approximately 91 % at all partial pressures of water. The following partial pressures of water were tested: $1*10^3$, $2*10^3$ and $3*10^3$ N*m^{-2}. In comparison to this for 1000 h an elimination rate of 99·9 % was achieved in the absence of water. The variation of mercury elimination capacity with different partial pressures of water is small. This behaviour may be explained by the formation of water clusters on the active sites.

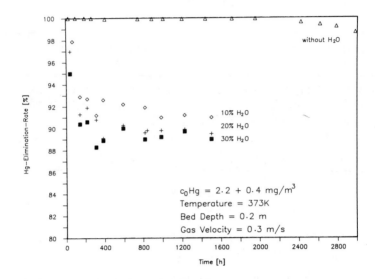

Figure 4: Effect of Steam on Mercury Elimination

4. Literature

/1/ Stock, A. (1934).
Angewandte Chemie 47, 64.

/2/ Pütter, K. E., Hirsch, M. (1934).
Angewandte Chemie 47, 184-185.

/3/ Matsumura, Y. (1974).
Atmospheric Environment 8, 1321-1327.

/4/ Showa Denke, K. K..
DOS 2 358 767

/5/ Krill, H., Wirth, H., Rittinger, G., Hohmann, V..
DOS 2 603 807

/6/ Sinha, R. K., Walker, P. L. (1972).
Carbon 10, 754-756.

A NEW TECHNIQUE FOR REDUCING HYDROCHLORIC ACID AND HEAVY METAL EMISSIONS RESULTING FROM HOUSEHOLD WASTE INCINERATION

J.P. Léglise - D. Abraham - D. Marchand - (SOGEA)

1 - FUME PURIFICATION: EXISTING PROCEDURES

Household waste incineration is a type of urban residue elimination system which has been in use for a long time: we can now reduce the residue into inert slag, while recovering large quantities of heat which can be used for the generation of electricity or urban heating.

The various national regulations limit the emissions of the main pollutants contained in incineration fumes, meaning mainly dusts, hydrochloric acid and heavy metals.

The regulations concerning dust emissions were implemented a rather long time ago, as dust removal techniques have been known for a long time.

Discharges from gaseous pollutants (in particular hydrochloric acid) have only been regulated for a few years.

Simultaneously, requirements were implemented concerning the limitation on heavy-metal emissions which, aside from a few exceptions (in particular mercury) are emitted in specific forms.

The changes of the regulations incited the development of numerous fume treatment processes, which can basically be grouped in three categories:

- the dry process, consisting in the injection of lime dust in the fumes, followed by dust removal by electrostatic filter or bag filter;

- the semi-dry process, consisting in the injection of milk of lime in the fumes, the evaporation of the water in the milk of lime, followed by dust removal by electrostatic filter or bag filter;

- the wet process, consisting in washing the fumes with water; the acid solution is neutralized and treated before being discharged.

These processes are characterized by generally satisfactory purification performances, but the highest performances are attained with the semi-dry or wet processes. However, all of these processes incur heavy investments and high operating costs (reagents, energy, maintenance), in particular for small- or average-capacity incinerators.

2 - A NEW PROCESS: DECHLORINATION BY CONDENSATION

SOGEA, with the participation of ANVAR* and AQA**, has developed a dechlorination process for fumes based on the use of a condensating heat exchanger (French patent no. 2 592 812).

The process comprises several successive stages which provide for:

- the trapping of dusts at the boiler outlet via an electrostatic filter.

Compliance with a residual dust content of 50 mg/Nm^3 ensures that the heavy metal particle content will be less than 5 mg/Nm^3.

- the trapping of the hydrochloric acid in a condensating heat exchanger.

The fumes are cooled to a temperature below their dewpoint temperature (approximately 50°C) using an indirect heat exchanger.

The incineration fumes of household wastes are very humid (10 to 20% in volume); it is therefore easy to condensate a large part of the water vapor which they contain.

The hydrochloric acid is absorbed by the film of condensed water which covers the walls of the exchange bundles.

---

* ANVAR: Agence Nationale pour la Valorisation de la Recherche (national agency for research valorization)
** AQA: Agence pour la Quality de l'Air (air quality agency)

The gaseous hydrochloric acid concentrations at the exchanger outlet are approximately several tens of mg per Nm³, for approximately 1 g at the inlet.

The acid condensates are recovered at the bottom of the exchanger, and a small part is carried by the fumes in the form of acid aerosol (approximately 100 mg HCl/Nm³).

- fume bubble removal: depending on the required discharge level, partial trapping of the aerosols can be carried out as a complement to the preceding stages, in order to obtain an overall discharge of less than 50 mg HCl/Nm³;

- condensate treatment: the acid condensates are neutralized with lime and decanted before being discharged.

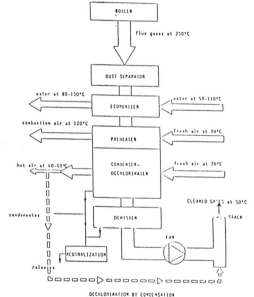

DECHLORINATION BY CONDENSATION

In addition to hydrochloric acid, other pollutants are trapped by the condensating heat exchanger:

- water-absorbable pollutants, mainly hydrofluoric acid (HF), which is trapped with a yield of 90%, and sulfur dioxide (SO_2) (trapping rates between 20 and 50%, depending on configuration retained);

- condensable pollutants, meaning 250°C gaseous pollutants, which condensate on the walls of the exchanger: this is the case for mercury (Hg), whose trapping yield exceeds 90%).

The condensating heat exchanger must be capable of operating under extreme conditions:

- temperature (250 to 300°C),
- acid condensates (in particular HCl and HF).

For this reason, the exchange bundles are made of materials which are both corrosion- and temperature-resistant, meaning in particular glass or graphite.

The exchanger casings are made of PTFE (fume side) or graphite.

3 - THE ADVANTAGES OF FUME PURIFICATION BY CONDENSATION

The main reason for the development of the SOGEA process was the reduction of fume dechlorination costs.

The heat recovered in the dechlorinating heat exchanger must therefore be generated in a useful form. In practice, the main way for valorizing the heat recovered is to increase the amount of vapor generated in the boiler: the process enables this.

The fumes are generated in the oven at a temperature of 1000° C, and their temperature is still between 220 and 250° C at the boiler-furnace unit outlet. They therefore constitute a significant recoverable energy source compared to the energy recovered in the boiler-furnace unit.

The use of this energy source is carried out in the three stages of the equipment:

Stage 1: Heating of the boiler supply water:

Heat valorized in the form of vapor

Stage 2: Heating of the oven oxidant air:

Heat not valorized in the form of vapor

Stage 3: Low-temperature heating of the air or water:

The boiler supply water is heated by approximately 30°C and the oxidant air to approximately 100°C.

Heat recovery in stages 1 and 2 is associated with a temperature drop of the fumes of approximately 100° C, the equivalent of approximately 10% of the heat received by the oven, which leads to a 10% increase in vapor production per ton of incinerated waste.

The heat recovered in stage 3 is at a low temperature (40 to 60°C). It can be in the form of warm water or air. In the latter case, this air can be mixed with the treated fumes, in order to desaturate them and resorb the visible wreath of smoke leaving the chimney stack. This low-temperature energy can also be used to heat the premises.

Main performance data of the process:

Purification

HCl residual content in the fumes: from < 50 mg/Nm^3
to < 250 mg/Nm^3
(depending on bubble removal)
Dusts: < 50 mg/Nm^3
Heavy metals: < 5 mg/Nm^3
Hg + Cd: < 0.2 mg/Nm^3

Energy recovery

Up to 350 kWh/t of waste (in the form of vapor)

Another advantage of the process, and not the least one, is that it is perfectly adapted for enabling conformity to the new standards of existing plants: as the condensating heat exchanger is located downstream of the dust remover, it can be installed upstream of a treatment line without making any major modifications to the existing plant.

4 - FROM EXPERIMENTATION TO THE FIRST INDUSTRIAL PRODUCTION

Following laboratory research, the process was developed using three successive pilot installations:

- a fumes/water prototype heat exchanger (1000-Nm^3/h output) enabled checking of the feasibility of the process and its sizing conditions;

- a fumes/air pilot heat exchanger (5000-Nm^3/h output) in continuous operation for more than 4,000 hours enabled more accurate sizing and the checking of clogging and corrosion resistance.

The pilot heat exchanger was constructed of glass tube exchange bundles and a PTFE-covered steel casing.

- a fumes/water pilot heat exchanger (5000-Nm^3/h output) entirely made of graphite (exchange bundle, tubular plates, casing) is currently in continuous operation (goal: 4,000 operating hours).

The measurement campaigns carried out on the three pilot installations enabled the development of a highly-sophisticated data processing model for the mass transfers and heat transfers within the heat exchanger which is applicable no matter what the cooling fluid used.

The process will be applied for the first time in an industrial context in the Rochefort-sur-Mer plant, which is scheduled to be commissioned in 1988.

This plant will be equipped with two 2.5-ton/hour furnaces and will treat the 23,000 tons/year of household waste generated by the equivalent of 80,000 inhabitants.

It will include two 15,000 Nm^3/h fume treatment lines and the heat exchangers selected will be of the fumes/water type, in graphite.

5 - CONCLUSION

- The implementation of new regulations for incinerating plants has led the concerned professionals and local organizations and authorities to question themselves concerning the additional costs imposed by an incineration system.

- Today, we can state that there are simple and reliable techniques which enable the financial expenses incurred by these new regulations to be limited, while maintaining high purification efficiency.

- With its high efficiency, its low cost price, and its capacity for standardizing existing plants, dechlorination by condensation will most probably be widely developed over the next few years.

HAZARDOUS SCRUBBER WASTE DISPOSAL AT VIENNA

MOSTBAUER P.
Institut für Wassergüte und Landschaftswasserbau
Technische Universität Wien, AUSTRIA

ZUSAMMENFASSUNG

Rauchgasreinigungsprodukte mit hohem Schadstoffanteil fallen in Wien bei drei Verbrennungsanlagen an. Die Menge dieser Sonderabfälle ist gering im Vergleich zur Menge der in diesen Anlagen verbrannten Abfälle. Die Zusammensetzung und Auslaugbarkeit dieser Abfälle wird kurz besprochen.

Derzeit ist keine Deponie für die Entsorgung verfügbar. Die Verfestigung wird als eine Lösung des derzeit anstehenden Problems betrachtet. Einen anderen Lösungsansatz bietet das von der Simmering-Graz-Pauker-AG entwickelte "Multi-Recyclo"-Verfahren.

ABSTRACT

Hazardous scrubber wastes are produced by air pollution prevention devices at three facilities at Vienna. The annual quantity of scrubber wastes is small compared with the amounts of wastes disposed of in these facilities. Composition and leachability has been described briefly.

At the present time no landfill site is available. Solidification seems to be a good solution for improving leaching characteristics. Another way is the "Multi-Recyclo"-process by which the generation of hazardous sludges can be reduced partially.

ORIGIN

At Vienna, three large waste incineration facilities are operating near densely populated areas. This is one of the reasons why air pollution control programs were established and implemented quickly. From 1985 to 1987 both facilities used in combusting municipal solid waste have been equipped with wet scrubbers of an improved LAB-EDV-type, a version of which had been installed before at Lyon/France.

Disposal of hazardous wastes and waste water treatment sludge is accomplished by two rotary kilns and two fluid bed ovens by one facility (EBS) located in a peripheral district of Vienna. A flue gas cleaning plant has started operation early this year and will bring about a tremendous reduction of acid gas emissions.

The annual amounts of scrubber wastes produced by these facilities are shown below:

FACILITY	FUEL	CAPACITY	INSTALLATION of SCRUBBER	ESTIMATED SCRUBBER WASTE QUANTITY
		t/a		t/a
Flötzersteig	municipal solid waste	180.000	1985	2.500
Spittelau	municipal solid waste	210.000	1986/1989	3.000
EBS	waste water treatment sludge	70.000 D.S.	1988	1.500
EBS	hazardous waste	100.000	1987/88	

Table 1: Incineration of wastes at Vienna

At the "Spittelau" facility there was an accident in the past year which damaged the flue gas cleaning equipment. This facility therefore will be out of operation till 1989.

COMPOSITION

The elemental composition of two samples taken from the "Spittelau" and "Flötzersteig" facility can be seen at table 2:

	Spittelau	Flötzersteig		Spittelau	Flötzersteig
water cont.	68.4 %	53.8 %	ppm Zn	19.000	11.200
carbon	2.8 %	1.1 %	ppm Pb	7.600	6.400
Ca as CaO	6.3 %	11.2 %	ppm Ni	120	65
K as K_2O	0.4 %	0.1 %	ppm Cu	700	700
Mg as MgO	2.6 %	2.4 %	ppm Cr	180	80
Fe	2.1 %	0.8 %	ppm Cd	240	230

Values except water content are based on dried samples (D.S.)

Table 2: Composition of scrubber wastes.

In addition to the above listed results scrubber wastes from waste combustion are known to contain mercury compounds, sodium salts, chlorinated dibenzodioxines and chlorinated dibenzofuranes.

It is remarkable that the main compound of the scrubber wastes described above is not calcium sulfite or gypsum, which could be expected when other fuels are used. This is due to the high ratio of fine particulates to sulfur dioxide of the flue gas coming out from the electrofilters.

At present no statistically significant concentrations of the compounds of the EBS scrubber waste can be reported. Gypsum and calcium sulfite content of those scrubber wastes are expected to be higher but composition will be fluctuating.

LANDFILL

In the past slag and ash from incineration have been disposed of together with municipal solid wastes without regard to possible **undesirable effects.** As a consequence, a landfill north of Vienna that had been used for dumping different wastes already 25 years ago had to be enclosed by impermeable walls to stop spreading of contaminants.

It is well known that compounds of heavy metals formed by precipitation with lime can be easily dissolved by acid leachate, e.g. leachate of municipal solid waste aged 1-2 years. (1,2) Experiments with electroplating wastes have demonstrated that the environmental impact of landfills which accept various types of wastes is strongly dependent on the sequence of deposition.(1) If scrubber wastes would be deposited below organic wastes a considerable subsurface transport of heavy metals can occur. On the other hand, putting scrubber wastes above municipal waste or waste treatment sludge brings about the danger of additional odor emissions. Sulfur compounds - mainly sulfate - are leached from scrubber wastes for a very long time and can be transformed biologically to H_2S in a reducing waste layer.

Fig. 1 demonstrates the leaching behaviour of sulfur compounds and chloride from "Spittelau" and "Flötzersteig" scrubber wastes. Conforming to the German standard leach test a water-to-waste ratio of 10:1

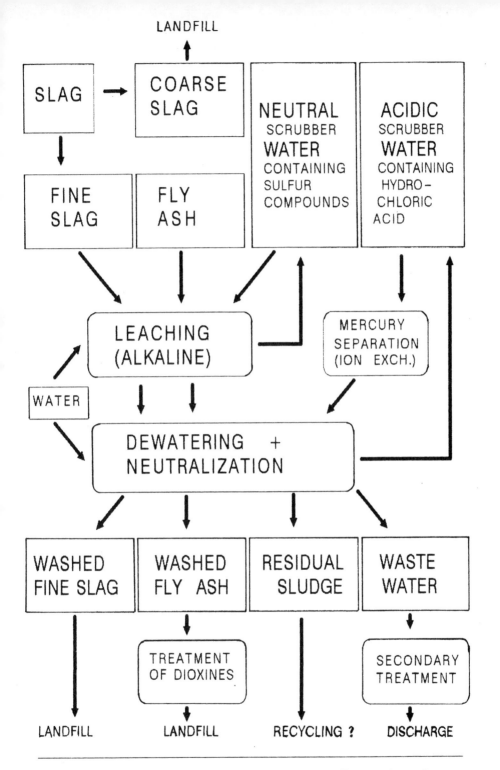

Fig. 2: Principles of the "Multi-Recyclo-" process.
WITH PERMISSION OF THE SIMMERING-GRAZ-PAUKER AG

has been used. Sulfur compounds have been determinated after elution by treatment with H_2O_2 and precipitation using barium chloride method.
Five successive elutions have been conducted. At first sulfur content of the filtrates slighly increases and then it remains nearly constant.

As risk assessment for codisposal landfill practices is very complex, in most countries as well as in Austria there is a remarkable trend to monofill landfill of wastes.

Although controlled monofill of hazardous scrubber wastes using double liner, leachate collection, treatment etc. ... would be a suitable disposal method until now no site has been available in Austria for Vienna`s hazardous scrubber waste. At the present time a part of it is exported and meanwhile the rest is stored under roof.

SOLIDIFICATION

Solidification is a method that significantly reduces the leachability of salts and heavy metal compounds by decreasing their mobility and the surface area of the waste exposed. Institutes at two Austrian universities have been commissioned to investigate the feasibility of scrubber waste solidification. (3)
At the present time (Apr. 1988) no results have been published.
To describe the advantages and disadvantages of solidification in brief we must say:
- Total weight is increased by solidification
- Test methods for assessing long-term behaviour are controversial

- Leaching of most contaminants is controlled by diffusion and therefore leaching rates will be low for a very long time
- Characteristics of solidified inorganic wastes are similar to that of a natural rock

RECENT DEVELOPMENTS

At present three residues are produced by the combustion of municipal solid wastes:

- slag
- fly ash (dust from electrofilter)
- scrubber wash water(s) or -slurries

As this slag is alkaline it should be possible to replace neutralizing agents like lime by the alkaline compounds of the slag. In fact, a pilot plant has been constructed by the Simmering-Graz-Pauker-AG (SGP) at the Flötzersteig facility using an aqueous extract of slag for neutralization of the acidic scrubber wash water (see Fig. 2).

The process has been named "Multi-Recyclo-"process and simultanously removes leachable heavy metals compounds from slag and fly ash. No hazardous fly ash or scrubber waste has to be disposed of since heavy metal compounds are concentrated up and treated separately. The amount of the residual sludge will be small compared with the amount of fly ash that is making diposal difficult today (4).

The next step in the development of scrubber wash water treatment systems will be the installation of a new scrubber including a version of the

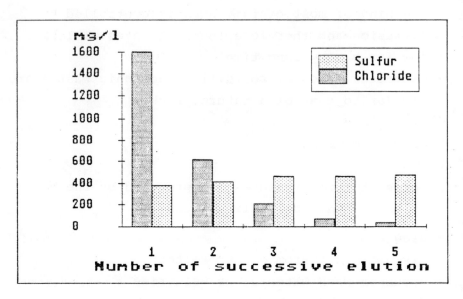

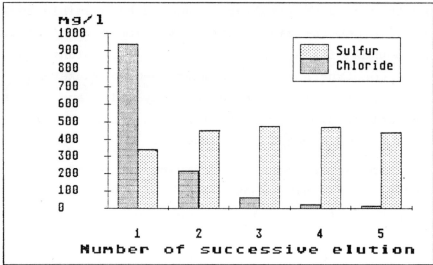

Fig.1: Leaching of scrubber filter cake from municipal solid waste incineration.

"Multi-Recyclo-"process at the Simmering facility. Our institute hopes to be able to report details about the leaching behaviour of the end products of this new process soon.

REFERENCES:

1. Mennerich, A. (1984). Gemeinsame Ablagerung von Hausmüll und Gewerbeabfällen. Müll und Abfall 8/84, 225-231.

2. Lechner, P. (1980). Sickerwasser bei gemeinsamer Ablagerung von Galvanikschlamm und Hausmüll - Lysimeterversuch. Wiener Mittlg. Wasser-Abwasser-Gewässer. Band 34, 01-028

3. Wruss, W. (1988). Personal communication.

4. Simmering-Graz-Pauker-AG. Das Multi-REcyclo-Verfahren (MR-Verfahren). (1987). Unpublished paper.

Experiences with Chlorine Balancing at an RDF Plant

Obermeier, T.; Freutsmiedl, L.

Presently cement kilns are the main customers for Refuse Derived Fuel (RDF) in the Federal Republic of Germany. The chlorine content of RDF often leads to problems during the burning process. Chlorine reduction measures in RDF plants should be based on a chlorine balance for the production process. This method was used at the RDF plant (ASA) in the Recovery Center Ruhr (RZR) in Herten, West Germany.

Sampling

Measuring points were defined at interesting units (MP 2 ... MP 6) and for the input and output streams (MP 1 and MP 7). Picture 1 shows the process flowline of the RDF plant in Herten with the measuring points: The refuse is handed from the feed hopper to a trommel screen (60 mm) by a plate conveyor. The last two divisions of the screen have big rectangular holes (300 x 700 mm). The oversize is bulky material (MP 3). The undersize < 60 mm is filled into containers, the middle fraction 260 mm > r_r > 60 mm goes to the hammermill (horizontal rotor with a 80 mm grid). A vertical air classifier/dryer separates the light and heavy fractions. In a second trommel screen (5 mm) the fines (MP 5) are rejected from the light fraction. The oversize (MP 6) is pulverised and pressed to pellets (MP 7).

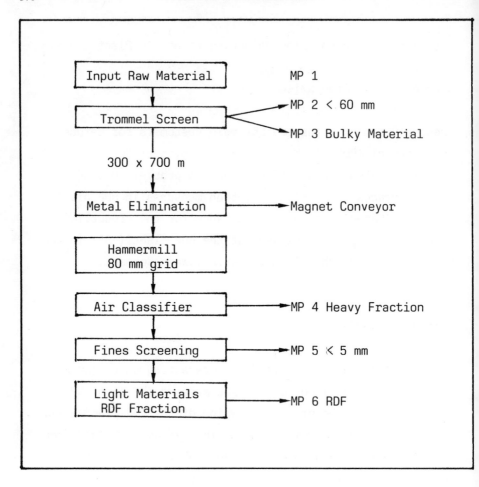

Picture 1: Process flowline RDF plant in Herten

At the measuring points samples were taken every 15 minutes during the 1,5 to 2,2 hour program (plus stops). The samples were mixed for further analysis handsorting (HS), sieve analysis (SA), ash content (AS), calorific value (H_o), water (WG) and chlorine content (Cl). Table 1 shows the chlorine specific analysis for the different measuring points. The 4 measuring programs varied in different input material from defined collection areas. During the 4 programs 240 Mg household refuse was processed, the sampling mass was 4 895 kg or 2 % of the whole input.

Measuring Point	Designation	Analysis
MP 1	Weighbridge	Input Weight
MP 2	Siftings < 60 mm	SA, WG, AS, H_o, Cl
MP 3	Bulky Material > 700 mm	WG, AS, H_o, Cl
MP 4	Heavy Fraction Air Classifier	SA, HS, WG, AS, H_o, Cl
MP 5	Siftings < 5 mm	SA, WG, AS, H_o, Cl
MP 6	Light Fraction Air Classifier	SA, HS, WG, AS, H_o, Cl

Table 1: Analysis for the different measuring points

Screen Analysis

A transportable trommel screen with exchangeable aperture was used at MP 2 (< 60 mm) and MP 5 (< 5 mm). From the light fraction (MP 6) and from the heavy fraction (MP 4) the portion < 8 mm was sieved. Handsorted plastic containers from the light fraction were also sieved (table 2).

MP	< 2	2-5	5-10	10-20	20-40	40-60	60-100	> 100
1								
2	x	x	x	x	x	x		
3								
4			< 8 mm					
5	x	x	> 5 mm					
6	x	x	5-8 mm	> 8 mm				
Ku-Fo								
MP 6					8-30	30-60	x	x

Table 2: Screen Analysis

Handsorting

The handsorting concentrated on the heavy fraction (MP 4) and the light fraction (MP 6). Table 3 shows the separated fractions and the used abbreviations.

Paper/Cardboard:	P/P
Plastic Foils	Fol
Plastic Containers	Foku
Textiles	Tex
Mixed Combustible	gbrb
Mixed not Combustible	gnbrb
Vegetables	Veg
Metals	FE/NFE

Table 3: Handsorting of the fractions at MP 4 and MP 6

Results

Picture 2 illustrates the chlorine distribution at the different output streams based on 100 % chlorine input. The chlorine contents at the measuring points are given in picture 3 (weight % dry mass). The mostly interesting light fraction materials are shown in picture 4.

It can be shown that at a chlorine content of 0,3 % based on dry mass 30 % of the input chlorine mass can be rejected with the undersize < 60 mm (MP 2). The portion of the undersize (MP 2) being 50 % wet mass of the refuse input, this chlorine load must be considered to be relatively small. The highest chlorine content is shown at measuring point 3 due to the high content of PVC-rugs. The relevance of this output stream (1 %

of the chlorine load) is low. 10 % of the chlorine load is
separated in the heavy fraction of the air classifier.
Screening the light fraction fines rejects 1 % of the chlorine
load. The biggest portion of the chlorine load is separated in
the light (RDF) fraction. An average amount of 60 % of the
incoming chlorine load is found in the light fraction, the
mass portion being only about 25 wet % of the input stream.
Picture 3 shows that the chlorine content of the light
fraction lies between 0,5 and 1 %. Picture 4 demonstrates that
plastic containers are the material group with the highest
single chlorine content.

As picture 5 shows, the total chlorine load of the RDF
fraction is determined to an extent of 35 % by the paper load.
Plastic containers are also important (28 %), because of
their high chlorine content ranging from 2,4 to 7,8 %, although
their weight portion is only 3 % of the light fraction. The
average chlorine content could only be reduced from 0,65 %
to 0,49 %, when all the plastic containers could be removed
from the RDF, the reasons being the influence of the foils
fraction and the relative high chlorine content of the paper/
cardboard fraction (0,3 to 0,6 %). The reduction of plastic
containers alone (e.g. by decreasing the classifying air
speed) can therefore not lead to an effective reduction of
the chlorine content in the RDF product.

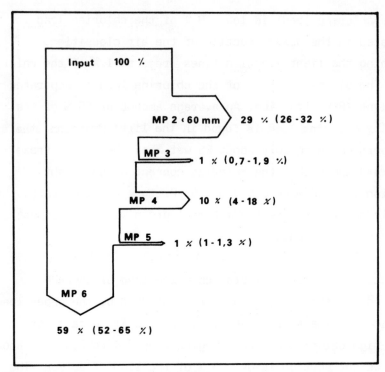

Picture 2: Chlorine distribution at the output streams

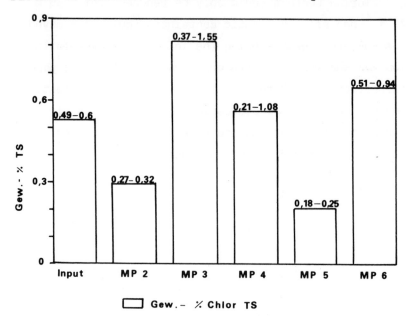

Picture 3: Chlorine Content at the different measuring points

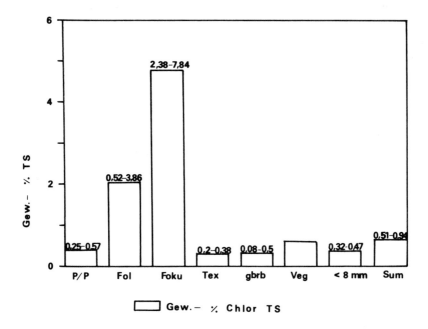

Picture 4: Chlorine content of different materials in the light fraction

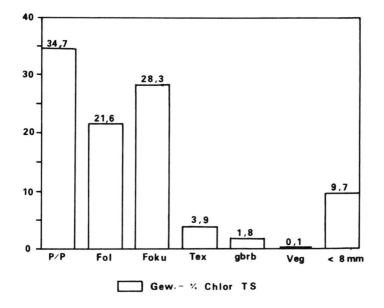

Picture 5: Influence of different materials on the RDF chlorine load based on the refuse mass stream

ENERGY VALORIZATION OF USED TYRES

Patrick Souet,
Assistant Director of Technical Action Services.
French Agency for Waste Recovery and Disposal

I. Processes Specifically Designed for the Treatment of Rubber

This category includes all the processes developed specially to solve the problem of old tyre disposal.

In order to be suitable for actual use by industry, these processes must successfully face up to two series of limitations:

a) technological limitations owing to the need for environmental protection and which will require carrying out a pilot project;

b) economic limitations: the size and operating costs (including redemption time of the initial investment) must correspond to the compromise between tonnage and buying costs able to be arrived at in the area.

I.1. Incineration

The processes presented in the following chart are incineration processes exclusively, which means the burning of the old tyres is complete in the combustion chamber, and the fumes produced are immediately cooled in the waste-heat boiler.

I.2. Pyrolysis

The processes presented in the following chart are pyrolysis processes exclusively, which means that the combustion of the used tyres is not completed in the combustion chamber, giving rise to the production of various combustibles, ranging from coal to natural gas.

I.3. Pyrolytic Processes Using After-Burning

In this chapter, two particular pyrolysis processes are presented. They are particular in as much as after-burning allows complete recuperation of energy simultaneously with pyrolysis. These processes are HOVAL and SIGOURE.

Process (Nationality)	Construction Design	Development	Production
HOVAL (Liechtenstein)	Hoval	France GENI-CLIMA FRAPY	Münchwillen, Switzerland 625 t/year
T.E.SIGOURE	T.E.SIGOURE		BELFORT-FRANCE

INCINERATION

Process (Nationality)	Construction Design	Development	Characteristics	Production
PYRALOX (France)	– AIR LIQUIDE – SEUM		Oxygen-Fed Rotary Furnace	PNEULAURENT Avallon France 1,500 t/y (1981) 3,000 t/y ('84) MICHELIN Spain 3,000 t/y (1984)
L. BOUILLET (France)	LAURENT BOUILLET INGENIERIE		Oscillating Furnace	Still being planned 1986
DEL MONEGO (Italy)	ENEAL			MARANGONI PNEUMATICI Rovereto, Italy 1984
VON ROLL (Switzerland)	VON ROLL	INOR (in France)	Pivoting Furnace	Pilot Project, Klus Switzerland
LUCAS (G.B.)	LUCAS FURNACE DEVELOPMENTS	LUCAS	Rotary Plate	GOODYEAR, USA District of Belfast, N. Ireland Avon, G.B.
GUMMI MAYER (W. Germany)	– GUMMI MAYER – UHDE	GUMMI MAYER	Furnace with oscillating Grate	GUMMI MAYER Landau, W. Germany 4,000 t/y (1973) 8,000 t/y (1983)
ECP (USA)	ENVIRONMENTAL CONTROL PRODUCTS			Old plans
FIRESTONE (USA)	BASIC ENVIRONMENTAL ENGINEERING	FIRESTONE		FIRESTONE – Des Moines, Iowa – Decatur, Illinois 1983
GOODRICH (USA)	CONSUMAT	GOODRICH	Combination of Household Refuse and Tyres	GOODRICH – Fort Wayne, Indiana – Tuscaloosa, Alabama – Miami, OH.

PYROLYSIS

Process (Country)	Construction Design	Development	Characteristics	Production
KOBE STEEL (Japan)	KOBE STEEL		Rotary Drum	Japan
NIPPON ZEON (Japan)	NIPPON ZEON		Fluidized Bed	Being tested
TOSCO II (USA)	TOSCO	GOODYEAR	Rotary Furnace, Steel Ball Heating	USA – Wyoming, Colorado
HERKO (W. GERM.)	HERKO-HERBOLD KIENER	MICLO (France)		
IRCHA (France)	IRCHA		Heat treatment with steam	In planning
KUTRIEB (USA)	KUTRIEB		Small dimensions	Pilot project, Wisconsin, USA
ENERCO (USA)	ENERGY CONVERSION Indiana USA			In planning
PTL (USA)	PTL TIRE SERVICE Ham Lake, Minnemata			In planning
DRP (W. Germ.)		University of Hamburg	Fluidized Bed	In planning
INTENCO (USA)	INTENCO	UGLAND, G.B. FIAT, Italy DECO, USA		

I.4. Depolymerisation: The UTC-IFP Process

This process was invented by the Université de Technologies de Compiègne and improved by the Institut Français de Petrole in collaboration with the Michelin Company and ANRED. It is still in an experimental phase.

The idea is to depolymerize rubber in a contact oil heated to 400°C. At the end of an eight-hour cycle, a hydrocarbon is obtained which is comparable to heavy fuel-oil n°2.

Approximately 600 litres of oil are needed to treat 200 kg of tyres to obtain 800 litres of fuel-oil.

This process has two advantages:
1) It transforms old tyres into a storable fuel which is suited to market requirements.
2) It can process tyres of all sizes, including those from civil engineering and public works, and the metal inserts are the only part in the rubber which needs to be eliminated **before treatment. The metal** parts can subsequently be reused.

This process works with different types of contact oils, but for economic reasons two would seem perhaps preferable: heavy fuel-oil and waste oil.

The low profitability of using fuel oil on one hand and the ban on burning waste oil on the other hand have hindered any industrial development for the time being.

II. Processes for Using Old Tyres in Existing Thermal Facilities

This category includes all processes which use an existing thermal facility in order to dispose of old tyres. These tyres thereby constitute a certain percentage of the heat load.

The operator of the facility is interested in:
a) The low cost of the tyre
b) The PCI of the tyre
c) The elements contained in the tyre (carbon, metals)

In order to be suitable for actual use by industry, it would be appropriate to verify through a series of tests that the use of tyres will not adversely affect the proper functioning of the facility or the quality of the effluents.

These processes are less vulnerable to supply difficulties. If the compromise between tonnage and buying costs is no longer practical, the facility is not threatened; the operator simply returns to the previous situation or seeks another alternative energy source. The investments involved are lower than in the first category since it is a matter of modifying equipment, mainly for handling the used tyres.

II.1 Incineration of tyres in cement plants

The cement industry is a tremendous consumer of energy. Producing cement clinker from limestone and clay requires a very high temperature: 1,800°C at its highest and an energy consumption of somewhere between 800 and 1200 th/clinker tonne, depending on the technology used. For example, cement plants with an annual production capacity of between 500,000 and 1,000,000 tonnes consume between 60,000 and 80,000 tep of energy per year.

The cement industry has been seeking alternative energy sources for many years. The process, requiring the most energy burns liquid wastes of both high and low calorific value.

II.2 Other Processes

Incineration of used tyres is also possible when mixed with:
1) coal
2) household wastes
3) foundry coke

CONCLUSION

Energy valorization processes of used tyres are separated into two broad categories:

a) Those processes specifically designed for treatment of rubber: 1) incineration, 2) pyrolysis, 3) pyrolytic processes using after burning and 4) depolymerization.

b) Processes for using used tyres in existing thermal facilities: 1) cement plants, 2) mixing with coal, 3) mixing with household refuse and 4) foundry cupola furnaces.

The feasibility of used tyre energy valorization depends both on the quality of the process under consideration and on **the feasibility of supply.**

In each of the two broad categories, there would seem to be only one process which has proved its worth: 1) incineration in the category of furnaces specifically designed for rubber and 2) cement plants for "processes using used tyres in existing thermal facilities".

INCINERATION IN CEMENT PLANTS

Country Firm	Location, Capacity (Clinker tonnes p/y)	Testing Began Investment	Tonnage of Tyres	Observations
Belgium CBR	Lixhe-les-Visé 1.5 million tonnes dry process		20,000 t/y of whole tyres 15% heat load	From 0.5 to 2 centimes/th of tyres
Austria	Gmunden 0.5 million tonnes	1981 – 2 million French Francs	2 tonnes/hour ground tyres (approx. 15,000 t/year)	Preparation cost of tyres: 400 FF/t (transportation and grinding) 7 cent./th
G.B. Blue Circle	West Bury		19,000 to 20,000 tyres/y 15% to 20% heat load	
Japan Nihon Cement	Saitawa 1.8 million tonnes		1.5 t/h, whole tyres 12,000 tonnes/year	
Canada Cement Lafarge	St.Constant (Quebec) 0.9 million tonnes per year	Experiments in 1982 and 1983 6.5 million FF	Potential capacity of 25,000 to 30,000 tonnes per year	Test results 30% acceptable substitution. Testing halting due to supply difficulties
France Lafarge	Lestaque (13) 0.35 million t	March 1982 1.5 million FF	5,000 tonnes/year whole tyres	6,000 tonnes/year capacity (15% to 20% of the heat load) Cement plant closed
Canada	Décaucour (Quebec)	1982 Testing		17% of the heat load

1) Furnaces specifically designed for incineration of rubber:.

It is to be noted that, in every case, the development of incineration processes which are actually put into use has been directly linked to a major brandname of the tyre industry, whether it be a manufacturer or a retreader. On one hand, this demonstrates the quality of those processes and on the other hand it proves that used tyres are not yet an alternative energy source suitable to the needs of communities (urban heating) or of industries other than the tyre industry. Certain incineration processes have proved their value, but the uncertainty of supply has been an obstacle to their development. Particularly since the idea of energy valorization has led non-industrial suppliers of old tyres to overestimate the value of their tyres, even if currently they are without other disposal outlets.

2) Use in cement plant furnaces:

Among the processes minimizing the risks due to the uncertainty of supply, only cement plants have stood the test of time. Moreover, it is to be noted that the tonnage in question is large and that used tyres from non-industrial activities are called for. Nevertheless, the obstacle to development remains the comparative cost in relation to other, cheaper alternative fuel sources derived from refuse and for which the cement plant can even -in extreme cases- be paid since they serve as a disposal outlet.

Session 9

Solid Waste Management in Developing Countries

SOLID WASTE MANAGEMENT IN METRO MANILA:
A SYSTEMS APPROACH

Melchor T. Gadi
Philippines

INTRODUCTION

Late in 1987, President Aquino created a task force to formulate a master plan to solve the solid waste problems of Metro Manila, a metropolis where more than 8 million people occupy a land area of 600 sq.km. Metro Manila generates about 3,500 tons of garbage daily. The bulk of these wastes are disposed in seven open dump sites while about 1/3 are dumped in creeks and rivers, burned or recycled. In the year 2000, the metropolis will have 10 million people producing 5,000 tons of solid waste daily. Clearly, something needs to be done.

Using systems approach, the task force was able to evolve a holistic solid waste plan which took into consideration the technical, economic, political, and socio-cultural constraints.

This paper will analyze the methodology employed in formulating the plan while at the same time describe the state of solid waste management in Metro Manila.

THE TASK FORCE : ITS METHODOLOGY

The task force is composed of ten government agencies: Metro Manila Commission, Development Bank of the Philippines, National Economic Development Authority, the departments of public works, environment and natural resources, health, transportation and communications, the Presidential Management Staff, and the local government of Manila.

In evolving the plan, a systems approach was used. The use of systems approach does not mean the application of complex mathematical models but rather the use of systems orientation in dealing with problems. This implies the recognition of the interdependence of all elements of the solid waste management system, the definition of alternatives and constraints, and the systematic evaluation of alternatives.

The management of Metro Manila's solid waste was viewed as an input-output system where the inputs are the wastes generated and the outputs are the final, processed waste disposed. To transform the inputs to desired outputs, several processes and operations need to take place.

INPUTS

Sources of wastes are households, street sweepings, commercial and industrial establishments. Per capita waste generation is about half kilo per day. Household wastes constitute 50% of waste, commercial wastes is the second largest.

PROCESSES:

Storage:

Various containers are used, ranging from plastic and paper bags to native baskets (kaing), cardboard cartons and sacks. Most are unsuitable for proper storage. In residential areas, it is a common practice to keep the receptacles inside the premises to avoid loss. They are only brought out during collection. Although there are communal containers to handle garbage in public areas such as markets, they are limited in number.

Collection:

About 70% of the wastes are collected. The collection process is simple, consisting of refuse pick-up by two or three crews at the street curb to a transporting vehicle and delivery to the disposal area. Street cleaners are employed to collect refuse materials from public areas.

Transportation:

Transport vehicles are hired from private contractors. These vehicles are inappropriate, they were originally designed for sand and gravel transport. There are about 400 of such open dump trucks.

Recycling/Resource Recovery

Recycling is already practiced to a considerable degree at the household level. Organic waste are utilized as feed for chicken, pigs and other household pets. A few households operate manual backyard composting. Compactor trucks have been acquired very recently.

To a certain extent, resource recovery is also practiced by the collection crew. They salvage the recylables during collection and sell them to junk dealers. This practice causes inefficiency, but it helps the collectors augment their low income (about US3/day). The recyclables are sold to numerous junk dealers which proliferate the metropolis.

Disposal:

Disposal is through open dumping. Although this method is simple and cheap, it is a serious threat to the health of the populace. One of the major dumpsites is 1.5 kilometer away from the source of water supply. Polluted water seeps into the water table and find their way into the major rivers.

There are seven dump sites in Metro Manila. The largest, called Smokey Mountain because of the illusion it creates, is in the heart of Manila. It takes about 1000 tons per day.

Burning is used to reduce the volume of garbage.

Administration and Finance:

Solid waste management for the metropolis is being

administered by the Environmental Sanitation Center. Resources are mostly allocated for collection, little is left for disposal.

The ESC's budget, which comes from the contributions of the cities that comprised it, is expended on the salaries and wages of 12,000 personnel. Of the total, 3,000 are garbage collectors while 7,000 are street sweepers and 2,000 are disposal area personnel. Payment for the hiring of dump trucks for collection slices a big chunk of its budget.

PROBLEMS AND CONSTRAINTS

To achieve the desired output of sanitarily disposed waste and recovered recyclable items, the following problems and constraints were addressed:

Political

The brief tenure of local officials, long-term career of bureaucrats, incompetent personnel, lack of initiative and motivation prevent the implementation of effective programs.

Traditionally, the attitudes of politicians and administrators is characterized by the "out-of-sight, out-of-mind syndrome". Any disposal method is acceptable, as long as the waste are hidden from the public's view.

Finance:

Solid waste service is underfinanced due to its low-priority compared to other public sectors. Most local officials view investments in solid waste management as "unproductive". It is more politically sound to spend on more visible infrastructures like school buildings or public markets.

Moreover, no direct charge is made for solid waste services. Hence, it has to compete for allocation with other services.

Public Attitude:

The most common attitude of public towards solid waste is that of apathy and indifference. The public usually associate solid waste management with inefficiency and inadequacy because most projects had failed to meet the public's expectations.

Socio-cultural

The dependence of the scavengers on waste as their only source of income is an important consideration in the implementation of any solid waste program.

Scavengers are mostly children and women who rummage through the garbage to retrieve resaleable items such as bottles, plastic, and paper. It is estimated that there are about 3,500 families with about 20,000 dependent engaged in scavenging in Smokey Mountain, Manila's largest dumpsite.

THE SOLID WASTE MASTER PLAN... LESSONS LEARNED

In formulating the master plan, the task force turned away from lofty objectives which often result in meaningless programs. Objectives that are specific, measurable and attainable were formulated. These objectives are easily convertible to work performance.

Based on a thorough analysis of the problem, several options were identified. These alternatives were pitted against a set of criteria which included: technological soundness and efficiency, affordability, ecological and environmental impact, and socio-political acceptability.

Although the primary considerations were economic and technical, factors such as economic dislocation of people affected, support of local politicians, were given weights in the final selection.

The final plan recommends a sanitary landfill with materials recovery and composting. Garbage collection in certain areas will be gathered in communal containers and hauled to transfer stations. Garbage collection in other areas will be undertaken by compactors and other dump trucks and transported to other stations. To optimize the resources available, efforts will be undertaken to maximize the effectivity and efficiency of the existing systems. Efforts will concentrate on improving the productivity of workers, streamlining procedures and improving the planning process.

Before implementing the plan, a grassroots public awareness campaign will be undertaken to demonstrate the benefits of the new system to the public.

Several important lessons were gained from the exercise. Most importantly, rigorous evaluation should be conducted on the appropriate technology needed. It should not only be assessed on technical soundness or economic viability but their adoption to local social and cultural environment. An excellent example of an inappropriate technology is incineration. It was ruled out as a viable option because the heat values of Manila solid waste are comparatively low. Moreover, incineration is a complicated technology requiring very high level of maintenance. What a Third World city needs are simple but effective systems with high reliability. An example is the composting system which utilizes a horizontally mounted rotating steel digestor cylinder which pulverizes and mixes raw refuse and at the same time conditions it for fermentation process. Systems such as this that are robustly designed and built to give high reliability under ardous service conditions should be the prerequisite of equipment and facilities in developing countries where low maintenance and high reliability are of utmost importance.

REFERENCES:

Metropolitan Manila Solid Waste Management Study, Quezon City, Philippines, 1982.

Solid Waste Management Plan for Metro Manila Task Force on Solid Waste Management, February 1988.

Gadi,M.T., Solid Waste Management in the Philippines, Ateneo Graduate School of Business, Makati, Philippines, 1986.

Session 10

Collection and Public Cleansing

Introduction of recycling based separated collection programs in relation to waste disposal methods.

Paul Masselink, The Netherlands

Introduction

Since 1979 rules for waste disposal have been enacted by law in the Netherlands, policy for waste disposal is based on: avoiding waste; reuse and recycling; **and disposal without damage** to the environment. Investments in environmental provisions have led to an increase of disposal costs, and through reduction of the number of dumping sites and plants to further centralisation. County Councils must design schemes to regulate location and way of disposal. By setting up these schemes it should be taken into consideration that future developments leading to a better waste disposal system must not be frustrated by current decisions.

Mechanical separation was used to gain resources, but quality standards were not met. Therefore collection of mixed resources separately was applied. Pilot projects showed that neither resources nor compost met the standard.

Collection of Vegetables, Fruit and Garden Waste.

In 1985 a pilot project, aimed at separate collection of vegetables, fruit and garden waste (V.F.G.), by weight the major component, was started with 1.200 households.

Project results were positive:
- the willingness to take part in source separation was high,
- the purity of the V.F.G. exceeded 95% on average,
- composting was easy, despite a moisture content of 60%,
- the compost was of high quality and met the severe standards to be introduced in 1992.

Heavy metals in V.F.G. compost.

	Cd	Cr	Cu	Ni	Pb	Zn
standards 1992	1,5	100	50	50	150	250
pilot project	< 1	36	39	12	62	179

With the response of 68% on average, 35% of the total amount was collected as V.F.G.. During the project, V.F.G. and remaining refuse were collected weekly. Opinion polls showed that up to 90% of the participants weekly disposed of all refuse. 75% of the participants were against a frequency of less than once a week.

Based on these findings the Province of Noord-Holland has decided to base her waste disposal scheme on separate collection and composting of V.F.G.

Collection models

In the Netherlands the collection costs are Hfl. 80,- per ton, which is equivalent to 60% of the total waste disposal costs. In the near future the costs for controlled tipping will rise to Hfl. 60,-/80,- per ton and the costs for incineration will rise to at least Hfl.120,- per ton.

Local governments made clear that increase of costs for the proposed waste disposal scheme must be limited to Hfl.20,-per ton.

Control of collection costs is possible using the current number of vehicles. One possible way is using a dustbin with two compartments. One disadvantage is a minor purity. This method also demands another type of collection vehicle and a reorganization of the schedules. Another disadvantage is the long-term inflexibility. Separation of a third component is only feasible against high costs and an **overall replacement of** equipment.

An alternative is to collect once every two weeks alternately V.F.G. and the remaining refuse with the conventional vehicles, without changing schedules or organisation. V.F.G. must be collected by a BIOCON. This method carries the risk of lesser co-operation because service is reduced and because of annoying smell during summer. It also has the disadvantage of inflexibility.

A collection scheme based on flexibility, current service level and cost control makes a new type of vehicle, collecting two components simultaneously, necessary.

A side-loader (K2-200) with 2 horizontal compartments (9 resp. 12 m^3) and separately regulated packing pressure has been designed. The net loading capacity is 12.500 kg.

Annual costs are low:

	content	investment	costs/year	c/hour
conventional	17m^3	Hfl.330.000,-	Hfl.106.900,-	Hfl.59,-
K2-200	21m^3	300.000,-	97.200,-	54,-

Collection Project.

It was decided costs of and attitude to 3 collection schedules were to be investigated.

Model A	V.F.G.	bin	120 ltr	weekly	K2-200
	rest	bin	120 ltr		
Model B	V.F.G.	biocon	80 ltr	two weekly	coll. vehicle
	rest	bin	240 ltr	two weekly	coll. vehicle
Model C	V.F.G.	bin	120 ltr	weekly	coll. vehicle
	rest	bin	240 ltr	two weekly	coll. vehicle

Beforehand the sanitation field has a strong preference for Model B, mainly based on cost calculations for the pilot area.

	bin	collection	disposal	total Hfl.
A	38	94	80	212
B	49	65	82	196
C	43	93	82	217

The models must be evaluated both economically and by non-monetary parameters such as flexibility and acceptance in view of future extending of separated collection.

Disposal and flexibility.
After V.F.G. collection the remaining refuse, increased in caloric value to 10,5 MJ/kg, will be incinerated in order to use the energy content. The caloric value is determined mainly by the amount of plastic and paper, of which plastic (7% in weight) attributes over 30%. Producing electricity in an incineration plant means that only 40% of the caloric value is gained. This means for plastic 17.400 MJ/ton and for paper 4.700 MJ/ton.
Using plastic waste for products replacing products made from virgin materials the net energy saving is 75.000 MJ/ton. Replacing wooden products energy savings are at least 26.000 MJ per ton. The net energy gaining making paperpulp from waste paper is 14.750 MJ/ton.

Technical developments allowing processing of contaminated and mixed plastic waste brings separated collection and recycling into reach. In principle all plastic can be reused; with paper about 50% is technically suited for reuse.
Incineration of plastic and paper as part of household waste in order to gain energy thus becomes a waste of energy as opposed to energy saving reached by reuse. Separated collection of plastic and paper (**response 60%**) **and** V.F.G. (**response 68%**) changes the composition of the rest considerably.

Composition of household waste

	1986	after separated collection of V.F.G.	V.F.G. & P./P.
V.F.G.	48,9	23,8	25,2
paper	23,8	36,6	30,5
plastic	7,1	10,9	8,0
glass	6,2	9,5	11,2
metals	3,3	5,0	6,0
textiles	2,3	3,5	4,1
	1000 kg	650 kg	550 kg
	7,5 MJ/kg	10,5 MJ/kg	9,0 MJ/kg

Three Components

In the near future the demand for waste paper will increase due to large scale investments in de-inking installations. Mixed plastics can be offered free of charge to recycling factories. Thus separate collection of paper and plastic is not only desirable but even profitable.

By choosing a collection model for V.F.G. the possibility of enlarging the number of separately collected components should be given prior consideration.

Collecting V.F.G. with the side-loader K2-200 answers this condition.

Collection scheme for V.F.G. and V.F.G. in combination with paper and plastic (P/P)

week 1	week 2	week 3	week 4	week 5	
V.F.G.	V.F.G.	V.F.G.	V.F.G.	V.F.G.	) K2-200
rest	rest	rest	rest	rest	)

week 1	week 2	week 3	week 4	week 5	
V.F.G.	V.F.G.	V.F.G.	V.F.G.	V.F.G.	) K2-200
P/P	rest	P/P	rest	P/P	)

Collecting a third component (e.g. paper/plastic) requires an extra bin, in which paper can be deposited. To avoid mixing paper and plastic, people must collect waste plastic in a plastic bag. On collecting day they put that bag in the extra bin on top of the paper.

Out of any 1000 kilos household waste 650 kilos remain after collecting V.F.G.. Are paper and plastic also collected only 550 kilos remain to be deposited of.

When planning new incineration capacity these tendencies should be taken into account.

Session 11

Resource Recovery and Recycling

WASTE REDUCTION THROUGH RECYCLING IN DENMARK YEAR 2000

Anton Elmlund, M.Sc.
gendan a/s, Copenhagen, Denmark

INTRODUCTION

My father, who was a seaman, used to say that once a course had been shaped it was technically easy enough to follow it and thus reach one's goal.

Already 15 years ago, gendan a/s shaped its course for the highest possible degree of recycling in Denmark! The technique that should make it possible to reach this goal is described on the following pages.

During recent years, a comprehensive legislation in the field of waste and recycling in Denmark has been created based on a large number of experiments, waste/recycling analyses, and discussions that took place during the seventies and eighties.

This paper examines the lines of the implemented, planned, and anticipated legislation and the consequences, especially the technical ones, these initiatives must be expected to have as regards Danish flow of waste and recycling in the year 2000. Furthermore, the paper shows the waste and recycling system that in gendan's judgement must be expected to predominate in year 2000. The system drawn up takes all examined legislative activities into account.

Apart from legislative tendencies, gendan puts forward a number of conceptual, behaviour, and market tendencies. These tendencies are the expedients that must be brought into play to reach the goal of the recycling course shaped.

The present paper contains, apart from this general introduction, the following sections, of which some are commented on, others are not:

1. Tendencies in recycling in Denmark.
2. Amounts of waste collected and recycled in Denmark in 2000.
3. Implemented, planned, and anticipated legislative initiatives.
4. Conceptual expectations.
5. The anticipated waste and recycling system in year 2000 and the flow of material for all Denmark.
6. Conclusions.

1. TENDENCIES IN RECYCLING IN DENMARK

The following tendencies can be deduced from developments during recent years.

A. <u>Tendencies in behaviour etc.</u>
   1. Generally, a growing motivation with the population and with enterprises to recycling and a willingness to cross familiar limits in terms of accepting a change of frequency, the bringing of the waste to the pavement, and the sorting into more fractions are seen.
   2. In future, a system of objectors will be a necessary element of any system of sorting at the source.

B. <u>Conceptual tendencies etc.</u>
   3. The future concept of waste will be:
      * materials (paper, plastic, plastic film, glass, metals, textiles,
      * compost,
      * problem waste,
      * energy, non-poisonous fuel (RDF),
      * dumping residue.

      A total of ten fractions. Sorting at the source will be unmistakably predominant.
   4. Basic sorting plants, i.e. relatively small, central plants with a purely mechanical, low technological and economical sorting technique producing homogeneous, coarse products, will be developed in supplement of the the sorting at the source.
   5. The conventional distinction between different types of waste will be replaced by a general view of the source. On a very long view only the type of material will be of interest.
   6. Bringing systems will more and more replace collecting systems.
   7. Voluntary organisations find much acceptance with the population, which fact should still be utilised to the benefit of collection at the source.
   8. In the near future, home composting will undergo an explosive development.

C. <u>Market tendencies etc.</u>
   9. Existing recyclers will continue to develop rational systems and to find new products and markets.
   10. Haulage contractors will become an important factor in the capacity of supervisors as well as motivators.
   11. The cost of alternative treatments (waste dumps, incineration plants etc.) will increase considerably.
   12. Generally, prices of conventional as well as unconventional return materials will be rising owing to
       - the development of new niche products,
       - better possibilities of local reuse, sale secured,
       - non-poisonous homogeneous fuel.

2. AMOUNTS OF WASTE AND REUSABLE MATERIALS IN DENMARK YEAR 2000

Year 2000 (1000 tons/year)	Private households		Trade etc.	Industry	Contractors
	Household waste	Garden and bulky waste	Commercial waste	Industrial waste	Building and demolition waste
waste food	390	(0)	120	165	(0)
cardboard	20	40	320	85	10
paper	285	125	115	195	(0)
o. compostible	255	370	55	30	(0)
plastic	85	30	40	40	10
o. combustible	35	185	50	90	415
glass	75	30	25	25	(0)
metals	50	60	30	330	(0)
o. noncombustible	50	55	25	405	2975
Total	1245	895	780	1365	3410

recycling amount with 1985-recycling level converted to year 2000

waste food	0	0	0	0	0
cardboard	0	0	90	30	0
paper	145	0	30	80	0
glass	50	0	0	0	0
other (metals etc.)	0	10	0	100	400
Total	195	10	120	210	400
recycling percentage	16	1	15	15	12

On the basis of a series of source-orientated analyses of waste and return products during the period 1978-88, gendan has determined the Danish flows of waste and reusable products. These analyses, which were financed by the Danish government, and an expectedly predictable development form the basis of the following calculations of a possible reduction of the amount of waste by the year 2000 by recycling. By-products, bottles for which a deposit has been paid, etc., are not included in the amounts stated above.

It is essential to have a basis as detailed as shown by the analyses above. Only an exact knowledge at a given time makes it possible to plan out different efforts and appraise their consequences. The reader is referred to: gendan a/s: "Frembragt affald år 2000", December 1987 (in Danish).

3. IMPLEMENTED, PLANNED, AND ANTICIPATED LEGISLATION IN DENMARK

Legislative activities	from-to year	active for these sources
1. Environmental protection law		
* waste/recycling planning	1985-1987	all sources
* waste tax to finance part of the recycling programme (40 DKK/ton)	1987	all sources
* collection of problem waste from private households	1988 (expected)	private households (collection from large sources have of course been running for several years)
* separate collection of garden waste and bulky waste. CFC tapping from refrigerators etc.	1990 (expected)	private households
2. Recycling law and dioxon action plan		
* special low-and-non-waste technology fund	1987-1990	industry
* recycling fund to promote R&D, plants, and special activities	1984	all sources
* collection by source of		
- newspapers and bottles	1990	private households
- paper from large offices	1988	large offices
- cardboard, paper, plastics	1990	trade
- waste food	1990	large kitchens

The above-mentioned legislative **initiatives** are already in operation or they are expected to come into operation. Apart from this legislation concerning waste and recycling there are laws concerning working environment, new limits to the emission from incineration plants, an environmental supporting scheme, and a special law concerning oily and chemical waste, and special waste from hospitals. The legislation in the field of recycling has been built up since 1978, before that time only the field of waste, in particular the scavenging, was covered.

4. CONCEPTUAL EXPECTATIONS

Year 2000	Collection system	Bringing system
time-independent system	*subdivided or multi-modular containers for cardboard, paper, plastic, and further waste from commercial and industrial enterprises and from contractors.	*compost containers with private households (garden waste and other compostible materials) *local recycling stations 1/250 households, only private persons admitted *central recycling stations (reception plants) 1/10,000 households, both private persons and enterprises admitted *bottle containers, centrally placed
time-dependent system	*system of bags/bins for further waste from private households *waste food will be collected separately three times a week from large kitchens *baling-press van will collect cardboard and paper from commercial and industrial enterprises, frequency-dependent *confidential and other kinds of paper from offices will be collected whole or destroyed *as is the case today, voluntary organisations will collect newspapers and glass from private households	*an "environment van", only for collections in connection with special campaigns for well-defined materials often topical in the current debate

As already mentioned, there are a number of trends of developments as regards the future concept of waste and reusable materials. As shown in the figure, these trends might f.inst. be comprised under: 1. type of collection, 2. type of frequency.

Generally speaking, there used to be bringing systems only for glass/bottles and bulky waste/garden waste, but gendan a/s expects a spectacular extension of bringing systems, with a relatively short distance to walk, however. A development will also take place of systems that aim at independence of time, i.e. systems that enable the citizens or the enterprises to decide for themselves when they want to deliver the waste. Whether the walking distance or the time independence is to be the predominant factor is at present being analysed by gendan in a series of projects. Provisional results will be available at the ISWA-conference.

At local stations the waste will be divided into nine types of material in subdivided special containers and minicontainers and in one maxicontainer for garden waste. Problem waste will be received too. At central recycling stations a division into 13 types of material in maxicontainers will take place and also a special reception of problem waste.

All remaining waste from private households and enterprises will be taken to a central low-technology, low-economy plant where an entirely mechanical sorting of the waste will take place into: * "non-poisonous" fuel, * potential raw compost, and * further waste to be dumped.

The ballistic separator will doubtless constitute the central unit of this basic plant. The "non-poisonous" fuel can either be burnt or it can be worked up further in a paper/plastic separation plant, but this will take place only regionally. Building and demolition waste will go to a crushing and sorting plant after having been sorted at the source, viz. at building sites etc., to secure that only clean rubble reaches the sorting plant. This plant crushes and separates magnetically the material and then turns out various **qualities of** graded gravel.

It is important that a standardised colour code is used at all sources so as to make the different types of material easily recognisable everywhere.

Finally, home composting will grip private households appreciably, both in self-contained and other dwelling houses. The proposed system for year 2000 as shown in the figure is based solely on gendan's estimate and takes all legislative activities into account. Accordingly, gendan's proposal may become a realistic system by year 2000. All elements of the recycling system presented have been tested in Denmark (1988) and the experiences obtained are utilised for further calculations.

5. THE ANTICIPATED WASTE AND RECYCLING SYSTEM YEAR 2000 AND A NATIONAL FLOW OF MATERIAL

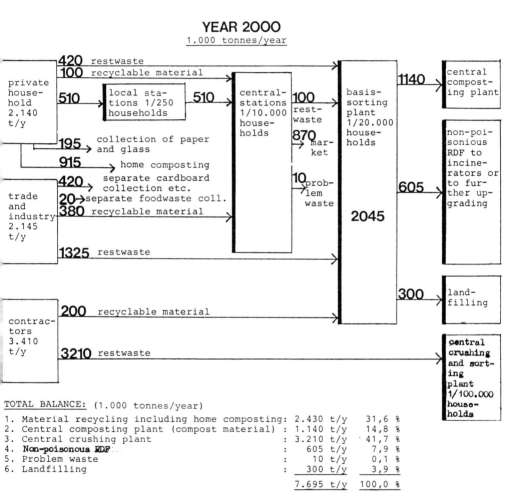

TOTAL BALANCE: (1.000 tonnes/year)
1. Material recycling including home composting: 2.430 t/y   31,6 %
2. Central composting plant (compost material) : 1.140 t/y   14,8 %
3. Central crushing plant   : 3.210 t/y   41,7 %
4. Non-poisonous RDF   :   605 t/y   7,9 %
5. Problem waste   :   10 t/y   0,1 %
6. Landfilling   :   300 t/y   3,9 %
                                                 7.695 t/y 100,0 %

On the basis of the assumptions made, the amounts shown, and the proposed system as per year 2000, it is now possible to calculate a flow of material for the year 2000. This will be seen from the figure. The flow was calculated for all Denmark. In the last section this calculation is followed up by some conclusions.

6. CONCLUSIONS

The calculations carried out lead to the following general conclusions:

1. It will be possible to redouble recycling including home composting in Denmark as a result of implemented, planned, and anticipated legislative initiatives. In addition to that, considerable amounts of reusable materials will be produced by central composting plants and crushing plants for building waste etc.

2. Problem waste will be removed from the flow to form the basis of a production at the contemplated basic plants of a non-poisonous fuel with a high effective calorific value.

3. If the proposed system is implemented in Denmark, the expenses to be born by the households will typically rise by approx. 50 per cent as compared with the present level of expenses.

   This calculation is based on the following estimate for 100,000 households:

400 local stations with containers	approx. DKK 60.0 mill.
10 central stations " "	approx. DKK 20.0 mill.
5 basic sorting plants	approx. DKK 50.0 mill.
5 central composting plants	approx. DKK 40.0 mill.
1 central crushing plant	approx. DKK 15.0 mill.
50,000 containers for home composting in self-contained houses	approx. DKK 30.0 mill.
5,000 containers for home composting in other dwelling houses	approx. DKK 25.0 mill.
100,000 bag/bin holders for all private households	approx. DKK 5.0 mill.
information of 100,000 households	approx. DKK 4.0 mill.
information of 1,000 enterprises	approx. DKK 0.1 mill.
Total investment	approx. DKK 249.1 mill.

   The crushing plant for building waste to an amount of approx. DKK 15 mill. should be paid for by the contractors, while the remaining investment should be divided equally among private households and commercial and industrial enterprises, which means an investment per household of approx. DKK 1,200, equivalent to a rise in the scavenging rate of approx. DKK 400/household/year, the operational expenditure notoriously corresponding to proceeds of sales in the system.

   Alternatively, the erection of another dump or incineration plant would today entail at least a doubling of the treatment expenditure as a consequence of, among other things, severer demands for environmental improving measures.

ELECTRO-OPTICAL SORTER FOR WASTE GLASS
Dipl.-Ing. A. Reichert
Prof. Dr.-Ing. H. Hoberg

1. Introduction

In the Federal Republic of Germany, a substantial increase of the recycling rate of waste glass can only be achieved by an improvement of the processing of cullet. In the production of green container glass the addition of cullet has already reached a maximum. However, yet much more waste glass could be added to the production of amber - and flint glass. This would require the provision of the respective quantities of glass sorted according to colours. There are basically two possibilities to accomplish this, a waste glass collection **with separate** sections for colours or a colour sorting which takes place after the glass collection.
The costs for the emptying of the bottle banks, the transfer and the transportation to the glass-smelter take the biggest part of the total costs of the recycling process. A separate collection raises the costs of exactly the before-mentioned steps considerably. Therefore, the colour-assorted waste glass collection will stick to its inferior role.

In the recycling of waste glass a colour sorting has as yet been carried out only by manual separation of entire bottles. This kind of separation is very costly and can only be applied to a fractional part of the total amount of waste glass.
For this reason, the Institute of Mineral Processing, Coking and Briquetting at the Aachen Technical University has developed an electro-optical sorter for waste glass cullet.

This sorter is able to make a colour separation of cullet even if the glass is soiled or partly **covered with labels.** A complicated and costly cleaning of the cullet can be dispensed with, which is a considerable advantage over any other test for the mechanical sorting of waste glass which has been carried out so far.

2. Sorting process

In the beginning of the seventies tests had already been carried out to sort waste glass mechanically according to colours. For this sake they used equipment which was usually employed for the colour sorting of mineral raw material. All such sorters work on the principle of reflection measuring, this means: the light reflected by the pieces is analyzed for the colour determination. These apparatuses turn out to be less qualified for the sorting of waste glass. On the one hand, this is due to the characteristics of waste glass and on the other hand, to the buyer's very high demands toward the assorted material.

Waste glass is usually extremely soiled. Besides, many pieces **are stuck with labels. Both facts lead to faults during the** colour determination by reflection measuring. For financial reasons, however, one has to dispense with a previous cleaning of the glass. Furthermore, the smooth, glossy surface of the glass produces unspecific reflexes which represent another cause of misplaced material.

Above all, the glass-smelters make extremely high demands on the colour-purity of cullet.

As an alternative for the reflexion method a colour determination by transmission measuring presents itself to glass being a transparent material. In the procedure applied here, white light is radiated through the glass fragments. The colour of the glass produces a characteristic course of the spectral transmission which forms the decisive criterion for the sorting. This curve is hardly even changed by colour-neutral soil. It is used for colour determination by a simultaneous measuring at several wave length ranges. During the measuring the cullet passes the colour sensor at a velocity of about 3.5 m/s. The measuring signals of the sensor are scanned with high frequency, so that even tiny fragments are analyzed in several points. In this way, even a colour determination of cullet which is partly **stuck** with labels becomes possible.

Throughout the development of the sorter, first of all a separation of flint - from non-flint fragments was realized to keep the mechanical-technical effort as low as possible. However, even a separation of amber cullet can be carried out without difficulties, as the spectral transmission curve of amber glass shows a rather characteristic shape.

3. Description of the sorter

The colour sorter illustrated in fig. 1 represents a testing equipment which should help to investigate the basic problems of a cullet sorting. It consists of two major functional groups, the mechanical part and the electric and electronic components.

The mechanical part has the task to separate the cullet, to let it pass along the colour sensor in an ordered way and defined position, and to catch the assorted pieces.

From a bin the cullet is passed on by a vibratory feeder to a multiple line vibratory conveyor. The latter consists of a vibratory chute which is covered by a serrate-shaped profile

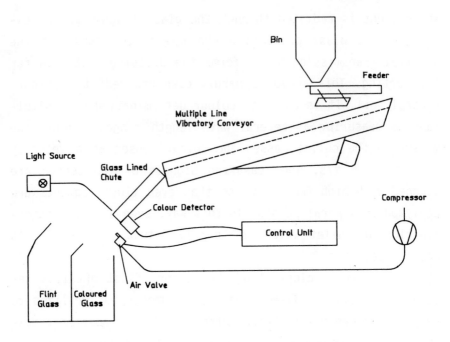

Figure 1: Outline of the Sorter

sheet-metal. The profile is shaped in a way that throughout the transportation the fragments are ordered one after the other. These grooves lead to more inclined slices on which the glass pieces are accelerated by gravity and therefore pulled apart. At the end of the slices the fragments pass the colour sensor. Shortly below, the pieces are deflected from their trajectory by an air valve with a flat jet nozzle and caught in a separate chamber.

The most important components of the electronic part are the light source, the colour probes, the amplifier, the digital analysis and control unit as well as the power supply for the valves.

The control system **consists** of modules: its basic set includes a processor board (Z80-CPU, 8 KB Eprom, 8 KB RAM, counter timer) and an interface board (A/D-D/A-convertor), which are fitted into a dust-proof enclosure.

It is to be noted that exclusively economy-priced and easily available standard elements are employed in the control system.

4. Technical data

The sorter is intended for a throughput of 200 kg/channel. The attainable product quality lies below 2 % extraneous cullet in flint glass. The recovery lies at 70 %.

The product quality depends highly on the throughput quantity: An overcharge of the vibratory conveyor is to be avoided by all means. Otherwise, a higher degree of negative recovery would be the consequence. Therefore, special attention has to be paid to the exact adjustment of the feeder.

The fraction size range of the glass pieces lies between 20 mm and 60 mm. Tests have shown that the material smaller than 20 mm leads to a heavy decrease of the throughput. Glass fragments of more than 60 mm size, however, cause disturbances on the vibratory conveyor, due to their unfavourable shape.

Raw material tests show that at this fraction size 45 % of the glass delivered at the West German waste glass treatment plants can be recycled by this sorter.

A technically and economically sensible size of a sorter lies at 16 channels. In this case, the average throughput is 3.2 t/h. The power consumption of such an equipment lies at 4.5 kW.

5. Possibilities of employment and costs

Principally, the sorter can be employed in any waste glass treatment plant. Apart from the described basic equipment the following additional devices would be necessary:
- belt conveyors
- two deck vibratory screen
- chutes
- product containers
- compressor

Till now, the electro-optical sorter has been built only as a test apparatus. Therefore, exact information about the price of the industrial version cannot be given.

For an apparatus of 16 channels including a compressor system, investment costs of about 120.000,-- DM have to be estimated.

Operating costs include energy consumption, staff costs and reparation costs.

The installed electric power is 19.5 kW including screen and compressor.

Staff is required only for an occasional control of the sorter, its maintenance and cleaning.

The wear of the mechanical part of the sorter lies in the usual scope which has to be expected of any such device. The electric unit contains only two elements with considerable wear: The valves (price: about 130.-- DM) have a durability of about 3,200 working hours. For halogen lamps (price: about 30.-- DM) the producer names a durability of 1,000 operating hours.

Session 12

Waste Reduction and Clean Technology

Investigations concerning the direct remelting of filter dusts from glass industry

Jürgen Kappel, Alfred Kaiser, and Dieter Sporn
Fraunhofer-Institut für Silicatforschung, Würzburg (FRG)

1. Introduction

Like other high temperature processes, glass melting causes airborne emissions of different pollutants. In waste gases from container glass furnaces the following contents were measured /1/ (in mg/m^3): dust = 103-356, SO_2 = 1172-2930, F^- = 2-30, Cl^- = 40-120. Since these values exceed the actual limits of clean air regulations in the FRG /2/ filter units with addition of basic sorbents for the removal of acid gaseous components have to be installed. However, a new environmental problem can occur since the disposal of the filter dusts is not yet clear.

Filter units in flue gas streams of glass furnaces with lime or lime milk addition collect about 2.5 to 3 kg dust per ton of glass production with the main components (in wt. %): $CaSO_4 \simeq 50$, $CaCO_3 \simeq 45$, $CaF_2 \simeq 1.5$, and $NaCl \simeq 2$. Additionally the dust may contain heavy metals originating from fuel or from impurities of raw materials. The situation in the lead glass and special glass industry, where those components may be the main constituents of filter dusts, will not be discussed in this paper but published elsewhere /3/.

Short term remelting of filter dusts in container glass works indicated no major problems. However, the question arises, whether long term accumulation of critical components will occur as it is known

from cement industry for thallium /4/. Calculations /5/ show that similar effects are possible at glass production. Some data are known from literature about limits of the incorporation of halides and heavy metals in different glasses and their influences on glass quality /6 - 9/, but generally the data are evaluated under conditions which are not transferable to industrial container glass production. Therefore investigations were started concerning the questions, whether
- direct remelting of filter dust as a batch component will be possible for long terms without affecting glass quality;
- filter dust alternativly can be re-used in other industries or for special purposes;
- conditioning of filter dusts, if necessary, is possible.

Filter units with lime or lime milk addition within the flue gas stream of two container glass tanks were investigated.

2. Results

Concentrations of critical heavy metals in filter dusts and in the produced glass were analyzed continuously. Figures 1 and 2 show the variations of mass concentrations observed within a period of 27 weeks when the filter dust was completely remelted in the same furnace where it came from. The batch composition was corrected with respect to the sodium, calcium, and sulphate contents of the dust. Most of the components have significantly higher concentrations in the dust than in the glass, but no accumulation occurred within this period. There is only a slight increase of the SeO_2 and V_2O_5 contents of the glass which takes place at a very low level and was not continued within the following time. Irregular changes of the concentrations of some components over an order of magnitude are not related to the filter dust recycling. The strongly varying SnO_2 contents are caused by the hot end finishing process of the glasses. The CoO ori-

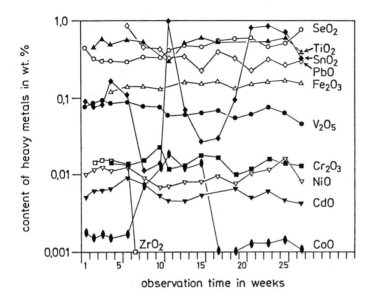

Figure 1. Content of heavy metals in filter dusts from a container glass tank

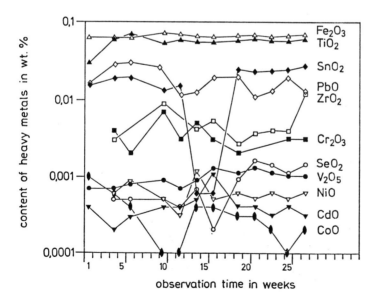

Figure 2. Content of heavy metals in container glass during complete recycling of filter dust

ginates from recycled cullets. Although only 10^{-4} wt. % of CoO can produce bluish colour tints /9/ these amounts do not cause any problems with glass quality. The high ZrO_2 content in the dust within the first weeks was due to repair work with ZrO_2 containing materials.

The contents of halides and sulphate in the filter dusts were determined (in wt. %) to: $Cl^- = 0.93$-1.44, $F^- = 0.60$-0.89, $SO_4^{2-} = 26$-32. Again there was no indication for accumulation processes. There is not enough knowledge about limits in Cl^- or F^- concentrations which influence the glass quality, the actual concentrations, however, can be estimated to be far below those limits. Since the batch was corrected with respect to the sulphate content of the dust also no problems due to that component occurred. These results indicate that even on long term remelting of filter dusts no problems have to be expected neither with glass quality nor with accumulation of critical components. The investigations will be continued until 1989.

In some cases, e. g. during periods of colour changing in a furnace or for special quality demands it might be necessary to collect the filter dust and to remelt it within shorter times in less sensitive glass tanks. Therefore one has to know whether it is possible to remelt higher amounts of filter dust. Laboratory experiments showed that sodium sulphate being used as a refining agent can be substituted in parts or even totally by the sulphate content of the filter dust originating from the fuel. Even an amount of filter dust doubling the sulphate content of the batch which is usually used is possible without affecting glass quality.

Due to the low specific weight (about 280 - 500 kg/m^3) of the filter dusts and to their tendency to cause secondary dust emissions their storage and handling might cause additional problems. Therefore experiments were carried out to investigate pelletizing of these filter dusts. With about 10 wt. % water pellets with diameters of 8 - 15 mm could be obtained without any problems. The pellets are mechanically stable and show only low abrasion during handling and transport.

3. Summary

Filter dust was directly remelted in a heavy oil fired container glass tank. During an observation period of 27 weeks no accumulation of heavy metals or other components neither in the dust nor in the glass was observed. Also the glass quality was not affected. The investigations will be continued until 1989. Also higher amounts of filter dust than those continuously collected can be introduced into the glass batch. This might be important in cases where filter dusts from several glass melting furnaces should be remelted within one furnace. In order to improve storage and handling properties the filter dusts can be pelletized without problems. This indicates that direct remelting of filter dusts from container glass production will be possible without causing secondary environmental problems. There is no need for re-using these filter dusts in other industrial processes. A waste deposition of the filter dusts will not be necessary.

4. References

/1/ Kircher, U. (1986). Fortschritte bei der Abgasreinigung von Glasschmelzöfen, Glastech. Ber. 59, 333-343.
/2/ Technical regulations for keeping the air clean: TA Luft vom 27. Februar 1986. Gem. Ministerialbl. A37 (1986) Nr. 7.
/3/ Roggendorf, H., Kappel, J., and Kaiser, A. Filter dusts from the lead glass industry in the Federal Republic of Germany, Glastech. Ber. to be published.

/4/ Welzel, K. and Winkler, H.D. (1981). Thalliumemissionen, Ursachen und interner Kreislauf, Zement, Kalk, Gips 34, 530-534.

/5/ Trier, W. (1987). Recycling of filter dust in glass melting furnaces. A theoretical study, Glastechn. Ber. 60, 225-233.

/6/ Hermann, W., Rau, H. and Ungelenk, J. (1985). Solubility and Diffusion of Chlorine in Silica Glass, Bunsenges. Phys. Chem. 89, 423-426.

/7/ Vogel, W. (1983). Glaschemie. 278. VEB Deutscher Verlag für Grundstoffindustrie, Leipzig.

/8/ Bamford, C. M. (1977). Colour Generation and Control in Glass, Glass Science and Technology, Vol. 2. Elsevier Scientific Publishing Company, Amsterdam, Oxford, and New York.

/9/ Merker, L. (1980). In Glastechnische Fabrikationsfehler (Ed. Jebsen-Marwedel, H. and Brückner, R.), 397. Springer-Verlag, Berlin, Heidelberg, New York.

5. Acknowledgements

The technical assistance of Messrs. Gerresheimer Glas AG, Düsseldorf (FRG), and Messrs. Oberland Glas AG, Bad Wurzach (FRG), and the financial support of the Bundesministerium für Forschung und Technologie are gratefully acknowledged.

INFORMATION SYSTEM FOR CLEANER TECHNOLOGIES
BY
JOHN KRYGER
CENTER OF CLEANER TECHNOLOGIES
TECHNOLOGICAL INSTITUTE
OF COPENHAGEN

1. Introduction.

The need for information on cleaner technologies, both generally and technically, is considerable in the industry and at the environmental authorities in Denmark. As a part of the Danish development program for cleaner technology, the Ministry of Environment through the Council of Recycling and Cleaner Technology have started a excensive informationcampaign. Apart from general information on cleaner technology the campaign includes research and development of a computerbased information system for cleaner technology(1)

Center of cleaner technologies (CCT) on Technological Institute of Copenhagen is in cooperation with other departments of the institute going to form app. 200 datasheets on cleaner technologies in the iron and steel industri in Denmark. Subsequently the useability of the datasheets are proved in selected industries in the branche or parts of it. To make the test effective the preparation of the datasheets will be concentrated on a few branches or industrial groups under the iron and steel branche. The project is financed by the development programme and

---

(1)   The development program on cleaner technology is financed by the danish government with app. 10 mio. $. The program is managed by the Ministry of Environment. The following branches are selected to research and development: Wood and furnishing industri, Iron and steel industri, Food industri, Electroplating industri and Tanning industry. Also substitution of chemical as CFC, mercury and cadmium is surported.

effectuated by the National Agency of Environmental Protection.

2. Intentions with the danish information system.

In 1985 the National Agency of Environmental Protection started a research on the knowledge and use of international catalogues and data bases for cleaner technologies in the Danish industry. The investigation is made by **CCT** and the report is finished in the summer 1988. The temporary conclusions have shown that the knowledge of catalogues and data bases is limited in the Danish industry. The ongoing enquiry to experts in the industry shall enligthen the question:"Are the informations in the catalogues and data bases usefull for the industri when chosing new technology ?" If not it will be necessary to prepare the datasheets in a new and more usefull form. The temporary results indicates that the latter will be the case for the majority of the Danish industries.

The aim in developing a database for cleaner technology in Denmark shall be seen in the ligth of this. The structure in the Danish industry with many small and medium-sized companies has shown that cleaner technologies cannot be implementated directly from foreing experiences. Actually these companies don't have the innovative power to work mission-orientated with the development of cleaner technolgies. Therefore is it necessary to adapt the international experiences and make an active marketing on these informations.

The information system is going to be a central on-line data base which also can be used by order inside some specific areas. The development and the maintenance is attended by a secretary assisted by some selected technical centers/specialists.

3. Description of the information system.

Administration, maintenance and marketing of an information database for cleaner technologies is sush a essential task that a central secratary is necessary. The secretary can be established in the National Agency for Environmental Protection or in an institute under the existing technological service system in Denmark for instance CCT on Technological Institute of Copenhagen.

The danish technological service system is an unique
advisory and development system established to support small
and medium-sized companies by lokal Technological Informations Centers assisted by som specialised research and
development centers in Copenhagen, Odense and Århus. The
technological service system are partly supported by the
National Agency of Industry and Trade.

CCT are convinced that the Danish service system (with
advantages) must play a central part in the development and
marketing of the information system. The technological
service system has great experience in consulting and have a
close contact to the small and medium-sized companies all
over Denmark. In the last 10 to 20 years the Technological
Institutes also had advised the central and local
authorities in technical questions.

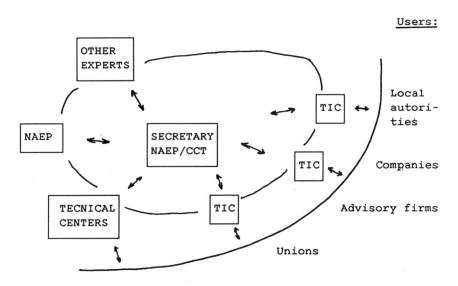

Figur 1: The information systems administative structure.
NAPE: National Agency of Environmental Protection.
TIC: Technological Information Centers.
CCT: Center of Cleaner Technologies.

As it is shown in Fig. 1. the information system administrative configuration is based on the technological service
system. The central secretary is assisted by some selected

technical centers at for instance the Technical University and Technological institutes. These centers are chosen on behalf of their technical knowledge and the relation to a given branch.

4. The contens of the data base.

CCT suggest that the data base should be structured at two levels. First level will contain short and appetizing informations about the cleaner technologies both technical, economical and environmental advantages and various references. The second level will contain a more accurate and detailed description and estimation of the technical, economical and environmental informations.

The two levels must contain:

Level 1	Level 2

Branche
Name of the process
Abstract
  - technical data
  - economic data
  - environmental data
Environmental efforts
Developer
References
Entry words
	Branche
	Name of the process
	Name of the existing proces
	The applicability of the process
	Technical data(1)
	Economic data
	Environmental data
	References
	Related datasheets
	Entry words

The second level is addressed to tecnical decisionsmakers in companies, in consulting firms and to the environmental autorities.

CCT finds that a databases on cleaner technologies will only find their use in smaller industries if they are heavily marketed by industrial consultans. Furthermore the database must include precise technical, ecnomical and environmental

---

(1)   The technical, economical and environmental data in this level are specified in proportion to the data in level one.

data, which can be recognized by industrial leaders. The combination of exact information and powerfull marketing will assure more efficient use of the informations in the database.

The existing databases covering the area are characterized by offering mainly general informations, and having drawn very little attention to themselves they are almost unknown to the the smaller and medium-sized part of the industry.

5. The State of the Danish Effort.

All activities - both the prelinimary investigations and the initiatives to build up the database - are financed by the development programme and effectuated by the National Agency of Environmental Protection. The table below shows the terms of building up the system.

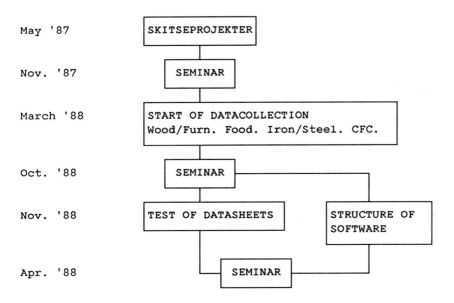

May '87	SKITSEPROJEKTER
Nov. '87	SEMINAR
March '88	START OF DATACOLLECTION Wood/Furn. Food. Iron/Steel. CFC.
Oct. '88	SEMINAR
Nov. '88	TEST OF DATASHEETS — STRUCTURE OF SOFTWARE
Apr. '88	SEMINAR

Parallel to this plan a minor computer based database is developed covering cleaner technologies in the electroplating industry. The experiences gained from this work are used also in building up the system.

The data-collection projects which are started in March 1988 covers the Iron and Steel industry, Wood and Furnishing industri and the Food industry. Under the Iron and Steel project substitution of CFC are included. These projects are ready in October 1988. After this a testing phase will be started. In this phase the datasheets are presented to selected factories from the branches in question to see if form and contents must be revised. At the same time the development of software will be started.

The construction of an information system as described demands large investments. The preparatory investigations alone cost app. 1 mio.US$ and a similar amount of money must be appropriated to include in the database the 6-8 most important branches in Denmark.

The remaining costs for building up secretariat, knowledgecenters and marketing are estimated to app. 500.000 US$ per year - i.e. when the system is fully built up(1) .

---

(1)   The financially estimations are made by CCT and are encumbered with some uncertainty.

Session 13

International Activities in Solid Waste Management

Session 14

Local, Regional and National Solid Waste Management

GENERAL PLAN OF SOLID WASTES TREATMENT IN PRAGUE

Tomas Bresitsláv, Praha

The development of the environment in Czechoslovakia represents one of the key problems of a nation-wide import, the solution of which is one of the most urgent objectives of the period till the year 2000. In connection with the improvement of the present living and working conditions it is also necessary to solve the problematics of solid wastes. Czechoslovakia as an industrial country produces a considerable amount of these wastes and their disposal is increasingly difficult. Solid wastes are especially solid municipal wastes, containing primarily solid household wastes, further industrial waste, building debris and other wastes, among which the most dangerous are those which contain matters endangering the health.

The problematics of solid wastes concerns most urgently the capital, for which Interprojekt Praha prepared according to the demands of the municipal authorities a general plan of solid wastes treatment. Interprojekt Praha has in this sphere considerable experience and it has been appointed the principal consulting organization in the sphere of treatment of solid municipal wastes in Czechoslovakia, whose task is to ensure the co-ordination of investment activities. The purpose of the General plan is to provide a basic conceptual guideline for the realisation of investments concerning the treatment of solid wastes in the Prague urban area. During the preparation of the general plan, it was necessary to take into consideration the following facts:

a) in Czechoslovakia, conditions necessary for the solution of the problematics of the solid wastes treatment on a desirable level are not yet satisfactory

b) not even in states with a high level of industrialization, where a great number of solid wastes treatment plants are already in operation, the opinions concerning this problem are not uniform and these very often antagonistic opinions become evident also in Czechoslovakia.

c) there does not exist any unproblematic process of solid wastes treatment

d) dumping must be restricted as much as possible, as there are no more feasible possibilities to dispose of solid wastes in landfills

The treatment of municipal wastes is controlled by local authorities. The treatment of other wastes, especially industrial wastes, is controlled by the ministries concerned. This applies also for Prague, which as the capital of Czechoslovakia, constitutes one of the ten counties of the country.

The capital, the city of Prague, about which, as the royal seat of Czech kings, the oldest document is from the 10th century, has at present about 1,2 millions inhabitants in an area of 495 km^2. The historic centre on an area of 9 km^2 is under care of historic monuments and it forms at the same time the basis of its present centre inhabited by about 70 000 inhabitants. Around the historic centre, the inner part of the city of about 80 km^2, formed by the former suburbs during the 19th and 20th centruy is situated.

In this area live more than 600 thousand people. The outer development zone of the city, whose number of inhabitants increased from the original 200 thousand to nearly 500 thousand, is formed besides the centres of the original suburban localities above all by housing and other buildings, constructed after 1945. Almost 40 % of the area inside the administrative boundaries of the city form the agricultural hinterland, the settlements of the border localities and other landscape elements. In this area, agricultural soil of good quality prevails, the preservation of which calls forth considerable problems connected with the solution of the requirements of the development of the capital, so that it is not envisaged that the increase in the number of inhabitants will exceed 1,3 millions inhabitants. The industry in the capital is considerable and it represents about 9 % of the production in Czechoslovakia. As regards the fuel medium, which considerably influences the composition of solid municipal wastes, it is assumed that in 1990 there will be used in Prague for heating purposes 50 % refined fuel, 40 % solid fuel and 10 % will be secured by district heating systems.

According to this prognosis, the following amounts are presumed:

in the year	1990	1995	2000
total amount of solid municipal wastes kt/year	415	430	445
specific amount of solid municipal wastes kg/inhabitant/year	352	365	370

According to the performed investigations, it is assessed that about 700 thousand m3 of industrial wastes and about 1,5 mil. m^3 building debris are produced yearly in Prague.

In Czechoslovakia, a special organization, Sběrné suroviny, carries out a separate collection in a well organized system of reception workplaces.

For the collection of wastes, collecting vessels of a volume of 110 and 1 100 l are used and it is not intended to use another system of collection in the future.

As for the installations for the liquidation of wastes, the first installation of this kind in Prague was the incinerator plant constructed in 1934. When choosing the site, already then the requirements of hygiene and economy have been taken into account. The plant is situated on the eastern part of the city and it is still in operation after manifold reconstructions. About 80 kt per year solid municipal wastes are being treated. In addition, in the 60th years a composting plant has been built in Prague, where approx. 40 kt wastes are treated every year. Agricultural establishments also collect wastes from restaurants and canteens for the production of fodder pastes. The remaining wastes are deposited in landfills.

As this situation with view to the entire lack of landfills in the capital is unmaintainable, Interprojekt proposed in the General plan successive conceptual solutions for the nearest period (till 1990), for the near perspective (till 2000) and a prognosis (till 2030) in the capital.

For the period of near perspective, the construction of a new incinerator with a capacity of 310 kt/year wastes is under projection. The construction of the new incinerating plant will be commenced this year on a site not far away from the existing plant, which will discontinue to operate after the new plant will be put into operation. For the city of Prague, a recycling plant of a capacity of 150 kt/year with a composting plant, delivered by the concern Vítkovice, is to be constructed according to the near perspective and a lesser one according to the prognosis.

The transport service of the capital which is responsible for the transport of collected solid wastes services 600 districts (urbans) from which wastes are transported. The transport is with the use of a computer system, organized so as to eliminate unnecessary losses during the transportation, and to ascertain optimum transport distances in relation to volume delivered.

In the General plan prepared by Interprojekt it was ascertained that till 2000 it will be necessary to invest about 9 milliards Kčs into the construction of new plants and their operation.

The proposed course of solving the conception of the treatment of solid wastes in Prague as projected by Interprojekt is prepared on the level of a general prospective plan. The succession and the system of equipment, operations and organizational measures are proposed in such a way as to take into account the present situation, real possibilities and development tendencies. The successive realization of the proposed partial constituents and of the whole system of solid wastes treatment in Prague depends on the realization of a whole range of measures on various levels and in various fields, and that is why measures proposed in the General plan are divided into fundamental and partial ones.

The most important information following from the General plan is the fact that the society cannot be satisfied only with the knowledge that the amount of wastes increases incessantly, but that it must try to eliminate the undesirable occurrence of wastes during the production, to restrict the introducing of secondary raw materials into the wastes and to eliminate from the wastes harmful matter which makes the subsequent treatment of wastes very difficult.

SOURCE SEPARATION OF MUNICIPAL SOLID WASTES
- FINNISH EXPERIENCES OF TWO CASE STUDIES
H. Juvonen [*], J. Kaila [**]

1. INTRODUCTION

The specific annual household solid waste generation in urban areas in Finland is 150-250 kg per capita. That figure includes separately collected materials like newspapers. The amount of household wastes actually disposed by the urban municipalities has been reduced by 10-20 % by weight by intensive household paper collection. This has affected significantly also the composition of household wastes disposed. Fractions like returned glass bottles and household machines and furniture are excluded from household waste generation estimates.

The essential features of household waste composition in Finland - compared with Central European or Nothern American countries - are:
1) the dominance of paper materials - newspapers, magazines and cardboard (45-55 w-%)
2) relatively high proportion of non-recoverable paper wastes (packages, soft household paper, diapers etc.)(10-20 w-%)
3) the minor proportion of glass and metal packages (5-10 w-%) due to effective return bottle system (90-95 % recovery rate) and use of fabricated cardboard packages laminated with plastics/aluminium.
4) lack of PVC- packages
5) relatively small amount of garden wastes due to climate and the dominance of living in blocks of flats in urban areas
6) relatively high proportion of kitchen wastes (25-35 w-%) due to habit of food preparation at home and lack of kitchen grinders (disposal by sewers)

---

[*] Technical Research Centre of Finland (VTT), Espoo, Finland
[**] KJA Oy, Waste Management Services, Helsinki, Finland

2. SOURCE SEPARATION OF HOUSEHOLD WASTES

Several source separation experiments have been accomplished in various industrialized countries in order to improve resource recovery and avoid environmental hazards. In Finland two source separation experiments were accomplished during 1984-88 in Helsinki metropolitan area (800 000 inhabitants) and the City of Turku (160 000 inhabitants). The goals of the projects were
1) determining the potentials for material recovery from municipal solid wastes and improving the quality of incinerator-bound solid wastes
2) estimating the actual participation of inhabitants
3) evaluating the costs and benefits of different separate collection and treatment systems

Several materials were sorted and collected during the projects. The framework of the projects is summarized in tables 1 and 2.

Table 1. Description of project frames.

A. **Helsinki project** (1984-85)
- organized and funded by the Council of Helsinki Metropolitan Area (YTV), Ministry of Environment and VTT
- project costs 1.5 million FIM - including the use of personnel and vehicles of YTV
- two subprojects (1. and 2.)

B. **Turku project** (1987-88)
- organized and funded by the City of Turku, Ministry of Environment and VTT
- project costs 1.0 million FIM - including the use of personnel and vehicles of City of Turku
- two subprojects (3. and 4.)

Table 2. Description of Subprojects:

---

1. PAPER & CARDBOARD
- effects on solid waste disposal
- traditions - separate collection began in the 1940's
- recovery system exists (Finnish pulp and paper industry)
- goal: determining the efficiency of separate bin collection from blocks of flats
- participants: 850 households in 12 block of flats and 2 terraced houses
- information: newsletters to each household

2. GLASS
- public and political demands
- collection and treatment system did not exist
- reuse system exists - quality demands are high
- goal: determining the technical, economic and sociological prerequisites for effective glass collection
- participants: 850 households in 12 blocks of flats and 2 terraced houses
  (separate collection bins at each house)
  4500 households in 5 residential areas
  (separate collection bins in each area)
- information: newsletters to each household
  guidance on collection bins

3. BATTERIES
- effects the emissions from solid waste disposal
- no collection and treatment system exists
- goal: determining the prerequisites for effective battery collection
- participants: full scale experiment (80 000 households)
- information: newspaper advertisements and articles, local radio programmes, information posters

4. METALS, ELECTRONIC EQUIPMENT, CLAY- AND CHINAWARE, GLASS, BATTERIES, THERMOMETERS AND GASEOUS DISCHARGE LAMPS
- effects the operation and emissions from the waste incineration plant
- no collection and treatment system exists
- goal: determining the prerequisites for segregative waste collection
- participants: 250 households in 10 blocks of flats
  (segregated collection bins at each block)
  5000 households in 6 residential areas
  (segregated collection bins in each area)
- information: newsletters to each household, guidance on collection bins, newspaper articles

3. EXPERIENCES OF SOURCE SEPARATION EXPERIMENTS

3.1 General

The results and experiences have been/ are being evaluated from the operational point of view of municipal solid waste management. The following conclusions are based on final reports of subprojects 1 and 2 (Juvonen & Kaila,1987 a) and intermediate report of subprojects 3 and 4 (Juvonen & Kaila, 1987 b). Key results are presented in table 3. The final reports will be published in 1988.

Table 3. Key results from source separation experiments.

Fraction	Subproject/ Collection system *)	Recovery rate (%)	Amount of recovered material (kg per capita)
Paper	1 / near	75 - 80	55 - 60
Glass	2 & 4 / near	30 - 50	3 - 7
	2 & 4 / area	10 - 30	1 - 5
Batteries	4 / near	> 50	0.10 - 0.15
	4 / area	< 50	0.05 - 0.15
	3 / town	< 25	0.02 - 0.10
Others	4 / near	< 25	0.5 - 1.0
	4 / area	< 25	0.5 - 1.5

*) Collection systems: near = bins at each block of flats
                                    area = bins in each residential area
                                    town = bin for each 1600 inhabitants

3.2 Effectiveness of paper separation in households

Recovery of household paper has been developed in Finland since the 1940's. Separate collection bins or containers at each block of flats have been used since the 1960's. During the experiment the existing collection schemes were changed and several newsletters were delivered to households in order to maximize the paper recovery rate. The achieved recovery rate (75 - 80 %) reduced the amount of household wastes 25 % by weight. It was estimated that by the existing collection schemes only 50 - 70 % recovery rates could be achieved.

3.3 Adoption of new source separation schemes

The results from new source separation schemes are not satisfactory in respect of recovery rates (table 3). However, there are several qualitative indications that during the 1-2 -year experiments only the most motivated inhabitants participated actively. It can be assumed that, for the majority, necessary changes of behaviour may take several years.

It was observed that the quality and quantity of recovered glass cullet depended on the collection system. The nearer collection points were situated, the better results were generally achieved. It was also observed that positive attitudes did not reflect actual behaviour.

A similar pattern has been observed with regard to source separation of batteries. A major positive attitude towards source separation of batteries was observed with subproject 4, but observed recovery rates were only satisfactory. In subproject 3 the increasing recovery results from the latter part of 1987 indicate that satisfactory recovery rates may be achieved in the near future.

3.4 Choice of information media

Information programmes are essential parts of source separation schemes. The information programme of subproject 2 was evaluated by the results of opinion surveys. It was observed that men and women had different patterns of following tv- and radio-programmes, newspaper articles concerning solid waste management and newsletters. Most women followed the newsletters more intensively than any other media. The pattern may reflect the fact that especially married women with children have no time for tv or newspapers.

3.5 Costs of source separation

The net costs of different source separation alternatives may vary significantly. In previous Finnish studies of separate paper collection it has been calculated that separate paper collection can be profitable. Based on the results of subproject 2 it was estimated that the net costs of effective separate collection of glass would be 200-400 FIM/ton. The preliminary net cost estimates for other collection schemes tested indicate 500-1000 FIM/ton net costs. These net costs may be considered rather high compared with the benefits.

References:

Juvonen H., Kaila J. (1987a). Source separation of municipal solid wastes. An experiment in Helsinki metropolitan area.
Source separation of glass. 61 pages.
Source separation of paper and cardboard. 46 pages.
Evaluation of household information programme. 40 pages.
Pääkaupunkiseudun yhteistyövaltuuskunta, Pääkaupunkiseudun julkaisusarja B 1987:1,2 and 5. Helsinki. (In Finnish, summary in Swedish).

Juvonen H., Kaila J. (1987 b). Study on solid waste management in the City of Turku. Intermediate report 30.4.1987. 46 pages. Technical Research Centre of Finland, Geotechnical laboratory. Espoo. (In Finnish).

R-98 PLANTS FOR RECOVERY OF RESOURCES BY MANUAL SORTING OF PRE-SORTED WASTE

Fritz Koza, I. Krüger AS, Gladsaxe, Denmark

1. INTRODUCTION

The Renholdningsselskabet af 1898, known as R-98, was founded in 1898 for the purpose of keeping Copenhagen clean. The company, which has the concession of transporting domestic waste, was put in charge of removing this as well as the industrial waste, etc. in the city of Copenhagen and employs a total of 550 persons.

Inspired by the growing recognition of the necessity of recuperating the resources hidden in our waste, R-98 decided to extend its activities to include the collection, treatment, and sale of resources for the recovery industries.

The company also acknowledged that economic aspects as well as the resources to be collected called for the development of an entirely new collection system, radically different from the previously employed principles. The main principle is that each individual waste producing unit split the recoverable resources into separate containers and keep them apart from the rest of the waste (including domestic waste).

The containers for the resources are specially designed to prevent the disposing of unwanted items.

The containers are emptied by means of compression vehicles which take the waste to the waste treatment plant described below.

2. THE COLLECTION SYSTEM.

The system, which is as yet in the start-up phase in companies and private households, has defined two primary material groups:

I. Pre-sorted material loads

II. Mixed material loads

The system further defines 5 material fractions:

1. Newspaper
2. Cardboard
3. Mixed paper
4. Plastics
5. Combustible material

The treatment plant is designed to treat pre-sorted as well as mixed material loads of fraction 1,2, and 3 and, as the case may be, fraction 4.

Group I: Pre-sorted material loads are defined as material loads consisting of the material fractions 1-4, however with the presence of other fractions to the extent accepted/defined by the costumer (e.g. a paper mill).

Group II: Mixed material loads are defined as material loads consisting of the material fractions 1-4, however with the presence of other fractions to the extent accepted/defined by the costumer and of fractions and quantities which are possible and/or economically feasible to eliminate by manual sorting.

Material loads not considered to fall under one of the two definitions will generally not be treated in the plant.

Primary design data:

Quantities:

I. Pre-sorted material loads	3.000 tons/year
II. Mixed material loads	18.000 tons/year
Total	21.000 tons/year

Material fractions:

1. Newspaper	15.000 tons/year
2. Cardboard	3.000 tons/year
3. Mixed paper	2.000 tons/year
4. Plastic(polyethylen) (not defined)	
5. Combustible	1.000 tons/year
	21.000 tons/year

Sorting capacity

The sorting capacity depends on a number of factors, however tests have shown that on an average one man can sort approx.   300 tons/year
(irrespective of material fraction)

Max. number of persons	8
Total sorting capacity: 8 x 300 =	2.400 tons/year

Demands to the materials to be sorted

Max. sorting capacity	2.400 tons/year
Max. permissible amount of other materials defined at 10%*	1.420 tons/year
100% pre-sorted ~	14.189 tons/year
Total	18.000 tons/year

Demands to the pre-sorting of the
material loads before manual
sorting ~ 75 %

* Depends on the specifications from each
 individual costumer

3. DESCRIPTION OF THE PLANT

The plant was installed in a large structure which was part of a major industrial plant built around 1960 for the manufacturing of the plastics polyethylene.

The design of the structure has heavily influenced the configuration and the building of the resource recuperation plant described below. These outer conditions have meant that the plant has become more complicated than would have been the case, had the structure been built for the present purpose.

The plant was built in two stages, and stage I was put into operation in February 1984. Stage I was intended for treating pre-sorted material, which meant that sorting should take place only occasionally and even then only to a very limited degree.

The annual amounts to be treated were defined at approx. 7.000 tons.

Stage II was put into operation in February 1988, and as previously mentioned is designed to treat pre-sorted as well as mixed materials.

The present plant comprises:
1. Receiving station incl. weigh bridge
2. Dumping area for:
2.1 Pre-sorted material loads
2.2 Mixed material loads
3. Conveyor line for:
3.1 Pre-sorted material loads
3.2 Mixed material loads
3.3 Pre-sorted material loads from silo and large items such as large cardboard boxes
4. Manual sorting section
5. Silo for:
5.1 Newspapers
5.2 Cardboard
5.3 Mixed paper
5.4 Plastics
5.5 Container for combustible material
6. Baling press
6.1 Press I
6.2 Press II
7. Storage for baled material

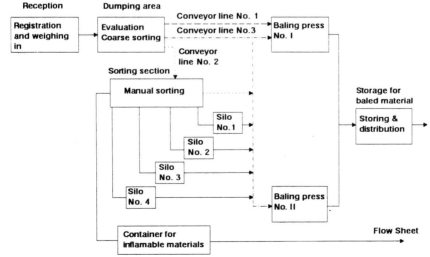

Flow Sheet

The annual amount being treated is as mentioned of 21.000 tons which can be increased to approx. 30.000 tons.

Plant Operation

Reception. The materials are delivered at the plant in compressing vehicles, weighed in at the reception.

Dumping Area. The registered materials are dumped **on to a** level floor in the dumping area for inspection. During the inspection the material standard is determined, e.g. if manual sorting is required.
According to the result of this evaluation, the materials **are loaded on to conveyor line 1, 2, or 3.**

Baling of pre-sorted materials. It should be mentioned that large unwanted items are removed from the material at this point. The pre-sorted material loads are taken on conveyor line 1 or 3 directly to the baling press, whereas the mixed material loads are taken to the manual sorting section on conveyor line 2.

Manual sorting. The manual sorting is carried out by a maximum of 8 persons, who sort from either side of a horizontal belt conveyor.

From mixed material loads consisting mainly of newspaper, the unacceptable amount of cardboard, other sorts of paper, plastics, and other types of combustibles will be removed. If the main substance is cardboard, the unacceptable amount of newspaper, other sorts of paper, plastics and other combustibles will be eliminated. This means that the persons eliminating cardboard in the first example will remove newspaper in the second.

The items eliminated from the conveyor are dropped through the floor of the sorting section into silos located below the floor, one for each class and one spare in case demands for an increase in the capacity should arise.

The silo capacity is ample in view of the manual sorting capacity, which means that they only need to be emptied once a day.

Emptying takes place e.g. after ending treatment of the last daily material load and one silo being emptied at a time (to avoid mixing the materials) by opening the side of the silo equipped with a material conveying bottom which leads the content of the silo to conveyor line No. 3.

Once one silo has been emptied, the second will be emptied and so on, until all silos have been emptied. Conveyor line No. 3 takes the material either to press I or to press II, as required.

Baling Press. The baling presses work continuously and automatically. The dimensions of the bales are h x b x l = 800 x 1000 x 1100 mm, each weighing approx. 300-500 kg.

Store. The bales are taken to the storage for baled material by means of fork-lift trucks equipped with a special bale device. The bales are stored according to material fraction in 5-bale high piles.

Storing capacity: approx. 6000 bales.
Storing period: 1-2 months

eba0408
EBA/ANG

CREATION OF A FACILITY FOR THE FABRICATION OF
SUBSTITUTE FUEL FROM INDUSTRIAL WASTE LIQUIDS,
SLUDGES AND SOUNDS PRODUCED BY EEC MEMBER COUNTRIES

M.Lamblin, D.Verbecke, Regional Council NORD, Lille
D.Lemarchand, JM.Tayon, SCORI S.A., Tourcoing

Over the past ten years, the French cement manufacturers (Société des Ciments Français and Ciments Lafarge) have developed the use of industrial wastes in cement plants. Together, the various studies undertaken, with the support of the French administration, have served to qualify cement kilns as a particulary effective tool for the destruction of industrial wastes. The technical quality of the elimination is complemented by the useful recovery of the energy content of the wastes.

Thus, the kilns of the French cement industry provide for the destruction and heat recovery of liquid wastes. In order to widen the range of incinerable products, SCORI, in cooperation with SCF and CL, has developed a technique for the fabrication of a combustible from wastes of various physical properties.

II. THE REGIONAL COUNCIL'S INTERVENTION

The COURRIERES power station (HBNPC) destroys about 100 000 tonnes wastes per year (from North of FRANCE and BENELUX).

The upcoming closure of this power station incited the regional council NORD - PAS-DE-CALAIS to subsidize a study about fabrication of substitute fuel from industrial wastes. The regional council operated with the French administrations the ANRED (Agence Nationale pour la récupération et l'élimination des déchets), the AFME (Agence Française pour la maîtrise de l'Energie), the AEAP (Agence de l'Eau ARTOIS-PICARDIE).

The study aims were :

- to treat and valorize the industrial wastes,
- to provide the regional cement activities with energy with an attractive cost price and to substitute imported fuel,
- to contribute to lessen the cost price of steam and hot water from the local heating stations.

The study was performed from March 1984 to July 1985 by SCORI S.A. with Société des Ciments Français, H.B.N.P.C., C.G.C., T.R.U.

The positive results of this study have led to the creation of two industrial wastes treatment facilities, one in BARLIN (North of FRANCE) and the other is SETE (South of FRANCE).

The regional council with the ANRED decided to subsidize the facility realisation, SCORI S.A. received in 1986 a special loan (880 00 FF).

III THE FACILITY FOR THE PRODUCTION OF SUBSTITUTE FUEL (COMBSU)

III.1 Definition of COMBSU :

COMBSU (substitute fuel) is a stable suspension with a lower heat value between 3 300 and 3 700 kcal/kg. It is composed of by-products and industrial wastes which can be liquids, solids, or sludges. It must be pumpable with a stability and homogeneity obtained by the combined effects of mechanical stirring and the chemical properties of dispersing agents.

From the point of view of the administration, COMBSU is considered a liquid hydrocarbon of category C 2, the same as the heavy fuel oils.

III.2 Characteristics and composition of COMBSU :

The characteristics of COMBSU presented hereafter have been specified in function of the needs and constraints of the users (cement manufacturers) :
. Lower heating value (kcal/kg) : .... 3 300 to 3 700
.   "   "   "   (kj/kg)   : .... 13 800 to 15 500
. Water (%) : .... 40 to 50
. Chlorine (%) : 1
. Sulfur (%) : .... 0,5 to 1
. Metals, Pb+Zn+Sn (ppm) : .... < 1 000
. Other elements, Cd+As+Cr+Hg (ppm) : .... <   300
. Flash point (°C) : .... >   55
. Viscosity (poise) : .... 2 to 4

At full term, the production on the present site is planned for 100 000 T/year.

III.3 The wastes

The wastes used, either liquid, solid or pasty, can be divided into the following categories :

AQUEOUS (Categories the most frequently encountered) :
. liquids contaminated with solvents (halogenated and non halogenated),
. oily emulsions and metal cutting fluid solutions,
. water from the washing of machined metal piece,
. mother solutions from fabrication (saline and non saline),
. residual mineral bases from chemical treatments,
. greasy water from the food processing industry.

DOPING PRODUCTS (Categories the most often encountered) :
. halogenated and non halogenated solvents,
. bottoms from the regeneration of solvents,
. whole oils from metal working,
. hydraulic transmission oils,
. non-chlorinated insulating oils,
. motors oils and mixed whole mineral oils,
. metal working sludges with hydrocarbure,
. liquid residues from distillation,
. pitch, tars, and asphalt,
. residues from decantation, filtration and centrifugation.

The parameters of products acceptability in the facility are the following :
. organic chlorine (%) : .... < 10
. metal - Pb+Zn+Sn - (%) : .... < 1
. other elements - Cd+As+Cr+Hg - (%) : .... < 0,1
. flash point (°C) : .... > 55

Note that these thresholds are higher than those for the COMBSU, a fact which must be taken into consideration during the preparation for production.

III.4 Description of the facility

The BARLIN facilities is composed by :

. weighing and storage facilities,
. storages of liquid raw material, sludges, and solids wastes,
. equipment for making COMBSU : a wash mill, a screen, a fuel grinding,
. fuel product storage.

The COMBSU facility is located in the Barlin cemetery and some equipments come from this cemetery.

III.5 Preparation of the stabilized mixture

The different steps which constitute the production process of COMBSU follows.

A) Preparation of the solid premix in the open pit :

The solid and pasty materials previously mentioned are premixed in a open pit tank in order to obtain suitable chemical and physical properties (handling and grinding).

B) Preparation of the liquid mixtures :

Prehomogenization of the aqueous liquids and doping agents is performed in the storage tanks. At this stage the fluidifying agents can be already incorporated into the composition, which in certain conditions permits an optimisation of their action by their prolonged presence in the liquids and by improved dispersion.

C) Feed to wash mill :

The premixed solid products are fed into the wash mill by repeated runs.

The premixing of the liquid products is achieved by pumping from the different reservoirs : aqueous liquids, doping or premixed liquids, liquid fluidifiers and surfactants.

D) Fabrication of the mixture :

The ultimate stage of the preparation of the mixture calls for the use of the wash mill, in which the liquids and solid phases are introduced in predetermined proportions.

When the stability criteria and homogeneity are correct, the COMBSU is sent to its final storage via a 2 mm filter.

It should be noted that this load can be reduced by proceeding by successive stages of enrichment of the COMBSU in solid matter. This type of manipulation uses an intermediate tank equipped for reflux operation which permits the improvement of the quality of the suspension.

During all these different steps, chemical and physical controls are performed on the COMBSU.

CONCLUSION :

The industrial wastes' treatment needs new processes and very high technical quality. The controls performed about incineration in cement kilns settle the high quality of the treatment. To widen the incinerable products, the "COMBSU" facility provides a medium calorific power fuel to the cement activity. This substitute fuel is a mixture of liquids, solids and pasty industrial wastes. This new facility answers the request of the industrials for the treatment of their wastes.

ABSTRACT

COMBINED ENERGY RECOVERY AND RECYCLING OF WASTES IN SMALLER REGIONS

Dr. Willibald Lutz, Vienna, Austria

Aside from the City and Province of Vienna the specific demographic and geographic situation of Austria requires smaller **catchment areas for waste treatment facilities. The annual** amount of municipal solid waste ranges from 177,5 to 385,9 kg with an average of 272 kg per head. A waste disposal plant **with the capacity of 100 to 300 tons per day may serve a po**pulation of 100.000 to 300.000 inhabitants.
The combination of energy recovery and recycling of valuable materials in a joint plant represents an optimal solution with regard to capital investment, operation and maintenance costs, environmental aspects, whereas reasonable tipping fees are the final result. A typical situation represents the City of Wels in the Province of Upper Austria.
In 1972 the municipality decided to install a mass burn incinerator with a capacity serving 100.000 people. After 15 years of operation the incineration plant has to be extended and refurbished. There are new environmental standards implemented in Austria regarding air emissions, water pollution and soil protection. In 1987 the municipality decided to shut down the old incinerator and build a new plant at the same site.
An international request for proposal was performed asking for a plant with the capacity of 8 tons per hour. The system consists of a presorting line, recovery and recycling of valuable materials, combustion, steam boiler, dust control device, wet flue gas scrubber, waste water treatment plant and energy recovery system by means of a steam turbine and generator. After a 7 months evaluation period the contractor was selected end of 1987. In 1988 the formation of the own and operator company, financing efforts and permitting is ongoing. The plant will be started up in 1992.
The capital investment amounts to AS 385 mill. including a secured landfill (multi barrier cell system) for the disposal of fly ashes and salt generated in the extended flue gas treatment device. The annual operation and maintenance costs including debt service amount to AS 63 mill. minus revenues of AS 29 mill. resulting average tipping fees in the range of AS 565.- per ton MSW.

COMBINED ENERGY RECOVERY AND RECYCLING OF WASTES IN SMALLER REGIONS

Dr. Willibald Lutz, Vienna, Austria

1. Introduction.

Austria consists of 9 provinces with a total number of population of 7,555.338 in 1981. There are 2.301 municipalities whereby the City of Vienna is a province as well. The quantities of municipal solid waste generated annually in each province are shown in table 1.

Table 1: Total quantity of municipal solid waste (MSW) in Austria in 1983.

Province	Population no.	MSW t/y	MSW/inhabitant kg/y
Burgenland	269.771	47.897	177,5
Kärnten	536.179	149.046	279,0
Niederösterreich	1427.849	275.164	192,7
Oberösterreich	1269.540	337.961	266,2
Salzburg	442.301	158.505	358,4
Steiermark	1186.525	240.429	202,6
Tirol	586.663	173.424	295,6
Vorarlberg	305.164	78.841	258,4
Wien	1531.346	590.992	385,9
Österreich	7555.338	2055.257	272,0

The amount of MSW per head ranges from 177,5 to 385,9 kg with an average of 272 kg per annum.
Because of the specific geographic situation of Austria in the Alps-Region the municipalities as well as the districts are rather small and transportation over longer distances is difficult. Therefore the catchment area of a waste disposal plant is limited to a population of 100.000 to 300.000 resulting in MSW-quantities of 100 to 300 tons per day.
Typical for such projects is the City and District of Wels in the Province of Oberösterreich (Upper Austria). In 1972 the municipality decided to install a mass burn incinerator with a capacity serving 100.000 people. After 15 years of operation the incineration plant has to be extended and equipped with additional flue gas treatment systems to meet the air

pollution standard of today. There also are new requirements regarding disposal of residuals from incinerators existing. An international tendering procedure in 1985/86 has shown that the modification and implementation of new flue gas scrubbers are more costly than the construction of a new single incineration line with the input of 5 tons per hour. On **this account the municipality decided to build a new plant** at the same site as the existing one, but with the capacity of 8 TPH.

2. Refuse incineration plant Wels.

In the City of Wels separate collection (curbside collection) of waste paper, glass and cans is ongoing since several years, resulting in a decrease of MSW in the range of 10 %. On the **other hand the commercial and bulky waste increases annually 5 to 10 %. In 1987 there was also an increase of household** refuse of approximately 5 %. Presently the total quantity amounts to 20.000 tons per year.

In the future the catchment area will be extended to the neighbouring districts with about 150.000 inhabitants. In addition to that there will be hauled residues with high heat value from other waste disposal plants such as refuse derived fuel, screenings from composting plants, biomass and others. The new plant design includes recycling devices to recover paper, plastics, scrap, aluminum, etc.

Table 2: Design parameters of the recycling and energy recovery plant Wels.

household refuse,	30.000 t/y
commercial waste, bulky waste, RDF, screening residues, biomass	30.000 t/y
total input,	60.000 t/y
average density,	130 kg/m³
moisture,	30 %
ash content,	25 %
heat value,	12,9 MJ/kg

In 1987 a public and international request for proposal was performed asking for a plant with the capacity of 8 tons per hour and alternatively 5,2 TPH. The system consists of a presorting line, recovery and recycling of valuable materials, combustion, steam boiler, dust control device, wet flue gas scrubber, waste water treatment plant and energy recovery system by means of a steam turbine and generator. Seven companies responded to the request for proposal quoting different

systems. The prices ranged from AS 250 to 350 mill. for the
8 TPH plant and AS 210 to 310 mill. for the 5,2 TPH plant.
6 companies offered grate type furnaces and one of them a
pyrolysis plant. In addition a fluidized bed incinerator, a
rotary kiln and a high temperature gasification plant was
presented. The tender evaluation was complicated and the
City of Wels asked two consultants for assistance. The evaluation was performed in three steps:
a) technical examination using a special system of 16 topics,
b) commercial examination including capital investment and
   operation and maintenance costs,
c) environmental analysis.
After a 7 months investigation period the contractor, a con-
sortium of the companies W+E Umwelttechnik AG, Voest Alpine
AG and Gerstl was selected end of 1987. In 1988 the formation
of the own and operator company, financing and permitting is
ongoing. The plant will be started up in 1992.

3. Plant description:

The refuse is delivered to the plant by garbage collecting
vehicles, container trucks, private vehicles or others. After
passing the weighbridge the refuse is dumped either into the
deep bunker for direct incineration or on the separate flat
storage bin for pretreatment and recycling. The burnable
waste is charged into the combustion chamber by a crab crane
and charging unit. According to the Austrian Standards a
minimum temperature of 900 °C of the combustion gases during
2 seconds has to be guaranteed. The slag is removed from the
grate by a grate siftings chain conveyor and the residue dis-
charger. A primary and secondary air fan supplies the required oxygen according to a computerized program. The hot
gases are utilized in the steam boiler. The high pressure
steam with a temperature of 400 °C and 40 bar is feeding a
condensate turbine connected to a generator with 6,5 MW capacity. This generator supplies the total energy of the
facility and the surplus is sold to the local utility.
The flue gas is cleaned in a multi step system. In the first
phase limewash is pumped into the reactor to remove gaseous
pollutants such as SO_2, HCl and HF. The reagents as well as
particulate matters are removed in the subsequent fabric
filter. Remaining small quantities of pollutants are eliminated in a multi stage wet scrubber using sodium hydroxide.
The slurry of the wet scrubber is pumped back into the reactor whereas the water is evaporated and the remaining salt
collected in the fabric filter. The ID-fan blows the clean
gas into the 50 m high stack.

LAY OUT OF THE RECYCLING AND INCINERATION PLANT WELS

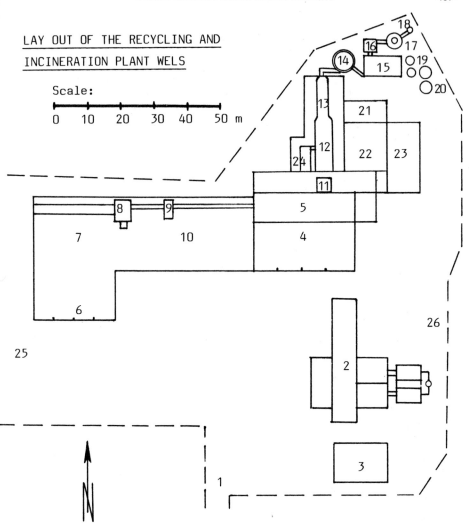

Legend:
1 entrance
2 existing incinerators
3 workshop
4 receiving hall
5 deep bunker
6 tipping area
7 flat storage bin
8 shredder
9 **scrap separation**
10 recycling hall
11 charging hopper
12 grate combustor
13 steam boiler
14 reactor
15 fabric filter
16 ID-fan
17 air scrubber
18 stack
19 lime silo
20 ash silo
21 pumping station
22 energy wing
23 cooling unit
24 hospital waste incinerator
25 residues disposal (secured landfill)
26 fence

The disposal of hospital waste is a problem in many parts of
the country. Therefore a small high temperature incinerator
will be attached to the combustor. A preliminary survey has
resulted in a quantity of infectious and pathogenous wastes
of 600 to 900 tons per year in the entire province. After
incineration in the combustion chamber the flue gasses are
led into the boiler and treatment system of the MSW burning
plant.
The slag is collected in special containers and trucked to
the neighbouring landfill. A private company removes the remaining iron scrap. Filter ashes as well as salt residues of
the flue gas cleaning are classified as hazardous and toxic
materials and have to be disposed in a secured landfill that
is permitted for such matters. A multi barrier cell system
will be implemented. The proposed quantity of slag amounts to
15.000 tons per year and ashes and salt will be 4.000 TPY.
The clean flue gas has to meet the following requirements.

Table 3: Austrian air pollution requirements.[+]

parameter:	mg/m^3 dry standard, 11 % O_2
1. particulate matter,	25
2. gaseous emission:	
hydrochloric acid, HCl	30
fluorine hydrogen, HF	0,7
sulphur dioxide, SO_2	100
carbon monoxide, CO	100
nitrogen oxide, NO_x (NO_2)	200
3. gaseous or solid emission:	
lead, zinc, chromium (Pb, Zn, Cr)	4
arsenic, cobalt, nickel (As, Co, Ni)	1
cadmium (Cd)	0,1
mercury (Hg)	0,1
4. organic substance (total carbon)	20
5. dioxines and furanes:	
2,3,7,8-TCDD 'äquivalent' ? Sense ?	0,1 ng/m^3
6. Temperature and time in the combustion chamber (flue gases):	
temperature,	900 °C minimum
time,	2 sec. minimum
7. **continuous measuring instrumentation:**	°C, SO_2, HCl, CO, CO_2, dust

* Note: Luftreinhaltegesetz - Kesselanlagen LRG-K 1988
(Regierungsvorlage 20. January 1988)

4. Costs:

The tipping fees of municipal solid waste as well as other residues depend on several facts. First of all the capital investment is important and furthermore the operation and maintenance costs and the revenues for recycling materials and energy have to be considered. Presently the market prices for recycled matters are low and sometimes do not cover the costs of the recycling process itself. Therefore the following calculation is based on zero revenues for recyclings.

Table 4: Capital investment, O & M costs and revenues.

1. Capital investment:	mill. AS
receiving and presorting,	26,3
grate and steam boiler,	73,1
steam turbine, generator, etc.	57,7
flue gas treatment,	88,9
electrical equipment, instrumentation,	31,8
civil works,	53,3
secured landfill,	20,0
site development,	14,0
engineering and permitting,	19,0
summary	384,1
2. Operation and maintenance costs:	mill. AS/year
personnel (34 men)	10,2
chemicals, water, residues disposal, etc.	15,4
maintenance and repairs,	5,9
insurance, tax, etc.	6,3
debt service,	25,6
summary	63,4
3. Revenues (electric energy),	29,5
4. Annual total costs,	33,9
average tipping fee,	(AS 565.-per ton MSW)

5. Summary:

Aside from the City and Province of Vienna the specific demographic and geographic situation of Austria requires smaller catchment areas for waste treatment facilities. The combination of energy recovery and recycling in a joint plant represents an optimal solution with regard to capital investment, O&M costs, environmental aspects, resulting in reasonable tipping fees as shown in the City of Wels, Upper Austria.

INTEGRATED REFUSE MANAGEMENT FROM A COMMUNITY VIEWPOINT:
THE CITY OF GRAZ, AUSTRIA

by Karl Niederl

1. INTRODUCTION

The exponential growth of the flow of energy and raw materials in the last three decades has lead to an avalanche-like increase of refuse in the areas of production and consumption. At the same time the quality of refuse has also changed - its " chemicalization " in all areas represents an ever growing ecological danger potential.

The only ecologically reasonable way to overcome this complex problem area proves to be the development of "integrated concepts of waste removal". The idea is that refuse should primarily be avoided and that unavoidable waste be recycled according to its components. Paper, glass, metal, etc. as valuable materials belong in the production cycle and the organic share in waste should be returned to the ground, the energy content of waste components with caloric value must be utilized.

2. PRINCIPLES AND GOALS FROM A COMMUNITY VIEWPOINT

2.1 Refuse Management and Avoidance Have to Begin in Production

Problematic materials once brought into circulation narrow limits are set for its utilization. Therefore production methods should be used, which manufacture products, which after usage or consumption can serve as raw material for the same or another product with no problem. But exactly the strong increase in "chemicalized" refuse contradicts this principle.

The community and federations can of course give certain impulses for avoiding and reducing refuse in the production area, but their influential possibilities are limited as long as there is no intervention in these areas by national laws. The realization of the refuse avoidance law being discussed in Austria would represent a first relevant step in this direction.

2.2. Avoiding Waste through Modified Consumer Behaviour

The decrease in the amounts of refuse and the reduction of the use of problematic materials in everyday living through goal-oriented and conscious consumer behavior on the part of the population represents the first step in the area of communal influence within the framework of integrated refuse management.

Commercial decisions are still felled primarily according to esthetic, prestige-oriented, and other aspects suggested by advertising often at the cost of the environment.

Therefore the awareness of the necessity and sense behind refuse avoidance strategies must be created in the population. This is where the communities can set the required discussion and awareness processes into motion by employing waste advisors for example.

2.3. Collecting Problem Waste as Prerequisite for Ecological Refuse Management

In order to reduce the pollutant potential of refuse and to allow for the ecological processing of waste it is necessary to <u>collect separately</u> as many <u>harmful substances</u> and pollutant products as possible before they are mixed with the remaining refuse (1). Once harmful substances are found in refuse they can be sorted out again technically in a very limited way.

2.4. Reprocessing of Substances by Means of Conversion into Garden Compost and Separate Collection Has Preference over Other Methods of Processing

The necessity of measures at the "source", e.g. in households, keeps increasing, the more consideration is given to saving resources and to the principle of foresight as determinate factors of refuse management. The separate collection of material and recycling of valuable or used material like paper, glass, metals, textiles, but also of waste that can be converted into compost, represents a relevant contribution to easing the burden on the ecological systems, because the flow of energy, raw materials, and pollutants can be reduced or interrupted by this.

2.5. Creating Incentives for Avoiding Refuse and Collecting Reprocessable Materials by Means of a Refuse Fee in Proportion to the Amount

Standard fees per household for the disposal of community refuse is still the most commonly used model of calculation. But this certainly does not present an incentive for avoiding waste, the separate collection of valuable materials, or making garden compost. To avoid refuse and to reward the utilization of materials the refuse fees should be calculated according to the amount of waste the garbage collectors pick up.

2.6. Methods for Handling Refuse Are Needed – Refuse Handling Systems Are to be Selected Such That Only Materials Fit for Final Dumping Result

In the future one will not only have to consider how to obtain as dense as possible a dump, but moreover through the employment of technological methods the originated refuse has to be transformed into an environmentally and for succeeding generations neutral state. One has to pay attention that substances dangerous to the environment are obtained in as concentrated a form as possible and substances compatible with the environment in as pure a form as possible, that is in an earthen similar form (2).

That is why natural organic composites should be re-used in as pure a form as possible, for example by making compost after separate collection of the organic waste, in contrast biologically alien organic connections like synthetics ought to be mineralized by means of a thermal procedure. The technical transformation has to take place in such a way so that the resulting emissions are kept at a minimum and remaining materials are fit for final dumping (2).

3. INTEGRATED REFUSE MANAGEMENT: THE CITY OF GRAZ

Already in 1983 the City of Graz decided on an integrated refuse management concept with the following items:

I. Provision to avoid refuse
II. Separate collection of different valuable material
III. Separate treatment of toxic components in refuse from households
IV. Conversion of waste to refuse-derived-fuel RDF and raw compost fraction
V. Processing of the raw compost fraction and sewage sludge into compost

VI. Energetic utilization of the RDF component
VII. Final dumping of the not recoverable residue, suitable from the environmental point of view.

3.1. Reduction of the Amount of Refuse in Graz

Did the amount of refuse to be disposed of in Graz increase by a yearly rate of nearly 7% to 95.000 t since 1965, a decrease could be registered since 1983 as opposed to many other communities, Fig.1.

3.2 Processing of Material through Separate Collection and Conversion to Garden Compost

According to experience gained from model experiments a collection system was chosen in Graz that is aligned to the potential amounts of valuable material in refuse.

The separate waste paper collection is conducted in a container designated especially for this purpose ("red container") immediately next to the container for household refuse. The collection of the remaining valuable materials colored glass, white glass, and metals are conducted according to a system of decentralized container set up.

By means of this system the fraction of the collected materials in the entire refuse already amounts to 24.7 percent, Fig.1, which corresponds with a doubling of the collected amounts of valuable materials as opposed to 1983.

The conversion into compost, especially in garden and residential districts is being developed and intensified with the assistance of the refuse advisers.

In order to obtain a better compost quality the separate storage in the so-called "bio-container" of biogenic matter from the common household is presently being studied and will be started in 1988 as a model experiment.

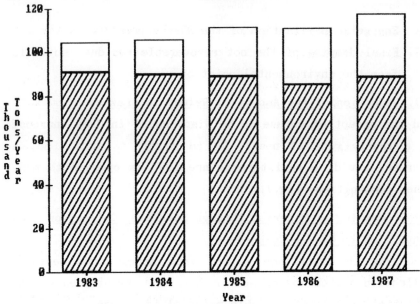

Fig.1   The development of refuse and collected valuable materials in Graz from 1983 - 1987.

3.3. Collection of Problematic Materials from Households

To reduce the content of toxic materials in households refuse a combined system of stationary and mobile collection locations were erected in Graz.

3.4. Processing of Refuse for Combustion and Compost

Goal in the processing of waste is the conversion of the delivered refuse, i.e. 88000 tons per year, into an as environmentally neutral condition as possible. The production and utilisation of RDF and compost from the bio-container among other possibilities is considered the most goaloriented means of processing refuse.

In order to realize this part of the Graz refuse management concept an agreement was made with VOEST-Alpine-Inc. This contract charges VOEST-Alpine Inc. with the construction of the waste processing plants and the responsibility of marketing and storage of the derived products.

3.4.1 Preparation of Refuse. In a refuse separation plant the delivered waste from Graz and vicinity is separated into the fractions: raw compost, RDF, hard matter, scrap iron and remaining material.

3.4.2 Thermal Use and Conversion to Compost on the Way to Realisation. It has not yet been possible to install the required compost plant as well as the RDF processing unit because the official authorizations for the originally decided locations could not be obtained.

The thermal processing of RDF should take place either in an industrial plant or to produce district heating in Graz on the location of the heat power plant. For the thermal processing of RDF (3), new techniques, for example high temperature gasification (4), (5), are to be introduced, capable of solving the main problems of thermal processing, like dioxines, heavy metals etc. far better than traditional combustion techniques.

4. SUMMARY AND PROSPECT

An integrated refuse management can only be intelligently based on the ranks of avoidance, reduction, reprocessing, and on elimination. Should avoidance not remain a mere slogan, frame conditions have to be provided beginning with production.

Recycling through measures at the "source", is superior to other methods of processing. Nevertheless, refuse processing methods are necessary, because it is ecologically not feasible that "only" 70 percent of refuse be deposited unprocessed (6).

It has to be considered that substances dangerous to the environment occur in the most concentrated form possible and substances that are compatible with the environment in the purest form possible.

The example of Graz shows that through measures by the community like refuse advice, collection of valuable material, conversion to garden compost, etc. the mountain of refuse can be reduced by one quarter. But it is also a goal in Graz to reprocess the amount of waste disposed of. The necessary units - compost plant and RDF processing - still wait to be errected.

5. REFERENCES

(1) W. Sattler, (1987), "Schadstoffreduktion im Abfall durch Maßnahmen der Verwaltung", Grazer Seminar Regionale Abfallwirtschaft, Techn. Univ. Graz

(2) P. Baccini, (1987), "Die Entsorgung in der Schweizerischen Volkswirtschaft - Leitbild 1986 - Prakt. Konsequenzen", Abfallbeseitigung - Erfahrungen International, Universität Graz

(3) H. Reimer, (1987), "Entwicklungsstand Verbrennungstechnologie Hausmüll", Abfallbeseitigung - Erfahrungen International, Universität Graz

(4) G. Staudinger, P. Freimann, (1987), "Vergasung von BRAM", Grazer Sem. Reg. Abfallwirtschaft, TU Graz

(5) P. Freimann et al., (1986), "The Voest-Alpine High-Temperature Gasification Process - a Contribution to Waste Treatment and Disposal", Recycling Int., Vol. 2, K.J. Thomé-Kozmensky Hrsg., Berlin

(6) O. Tabasaran, (1987), "Abfallwirtschaft - Stand und Ausblick", Grazer Sem. Reg. Abfallwirtschaft, TU Graz

SOLID WASTES MANAGEMENT IN RHODOS

E. Papachristou, Aristoteles University Thessaloniki, Greece.

Introduction

Over the last few years, a strong concern has been developed in our country, regarding matters of environment and quality of living.

The protection of the environment from human activity's effects is definitely a universally admitted necessity. One of the most important negative interventions of man towards his exterior environment is the un-controlled disposal of the solid waste he "produces" which bears dangers against his health, directly and in the future.

The need for a correct and controlled disposal of every useless product to-be-refused is urgent, because waste constitutes a collection of the most heterogeneous materials, with worrying quantities of toxic substances and pathogenic sites, that are in immediate threat to people and environment. Looking at the subject from an ecological, hygienic and aesthetic point of view, we're driven to the conclusion that we owe our respect to the environment around us, which, even now, is being **exploited** and destructed by man's ignorance.

In the developing countries, technology is advancing towards the stage of the **exploitation of solid** waste. Recycling, Energy safe (combustion, pyrolysis) and Compost production are, since long ago, part of the productive procedure applied in the Sector of environmental technique that, along with the continuous improvements it brings about in the method of sanitary burial, succeeds in reducing to a minimum the natural environment's deterioration.

The Problem in Rhodos

The island Rhodos and the greater coastal area are considered very important assets of this part of the Mediterranean Sea. Located at the South-East side of the Aegean Sea, the island Rhodos is actually used by various activities of economic and social interest and mainly for recreation. Its permanent population is about 40.000 but during the summer, because of the foreign tourists, the population becomes about 300.000. Thus is obliged to protect the beach, the underground water, the forest and rather little land.

The waste's composition plays a major role in choosing the method of disposal. The study of the waste's composition and physical-chemical properties constitutes the base of every management and disposal planning.

The Municipality of Rhodos considering the above thought a basic principle, charged the Departement of Hydraulics and Environmental Engineering, with the research of the qualitative and quantitative analysis of the city's solid waste.

Results and Discussion

The city of Rhodos is divided in nine parts, that

SOLID WASTES MANAGEMENT

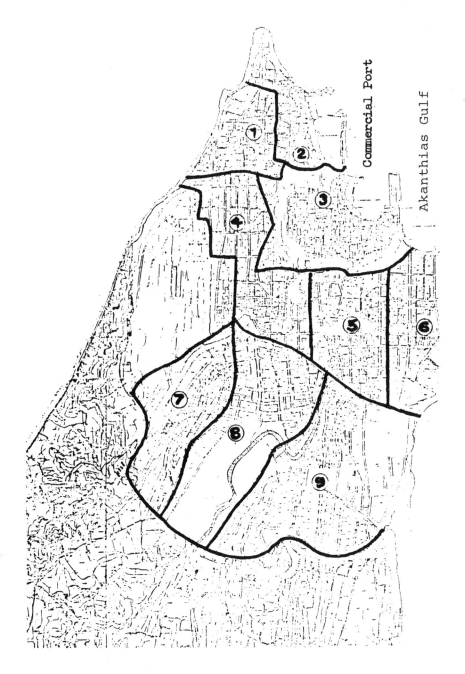

Rhodos, City and Greater Area

contain three building squares. The sample collection occurs in the morning every day since 4.9.1987 The project contains sample collection from all the parts of the city (industrial area, tourist zone, center, ancient town, urban area) for all over the seasons of one year (1). The sample volume is 1.1 m^3. This sample is weighted and divided proportionally to the kind of substances. Thus we have the following categories: organic substances, paper, plastics, rubber-wood-leather-cloth, metal, inert.

The material categories are weighed and their percentage in the waste is determined on the spot. After the weighing procedure is finished, metals, glass and inert materials are removed, while the foodwastes are cut in pieces and mixed. The final 1,5 kg sample is first desiccated at 105°C for the determination of volatiles, and then homogenized in a grinding mill into grains of 0,5 mm's diameter. A quantity of the ground desiccated sample is put into an oven where, after the combustion, the ashes' percentage is determined. The already removed, during the sorting-out, materials, i.e. metals, glass and inerts, are also taken into account in this percentage.

With the proper conversions, the percentage of the combustible materials contained in each liquid sample's kilogram of solid waste is then deduced by calculations.

The calorific value is determined on ground desiccated sample, in a Galennkamp type adiabatic calorimeter, with the ASTM/D 240 method.

In order to receive the heavy metals (Fe, Cu, Pb, Cd, Ni) out of the fermentable materials, we use the method of extraction by aqua regia (HCI: HNO_3 3:1) while in order to determine them we use an **Atomic Absorption Spectrophotometer equipped with Graphite Furnace** (Perkin-Elmer Mod 372, HGA 300).

Finally, an automatic Perkin-Elmer 240-B analyser is used for the determination of total carbon and **total nitrogen in the fermentable materials.**

The international bibliography mentions that in order to determinate the calorific value of waste, it is necessary to know its capacity in ashes, volatiles and combustible materials. These elements will constitute a threeagle system of factors, known as the TANNER diagram (2), which will contribute to the determination of the self-perservation combustion area i.e. the conditions under which a material can be burned without the addition of any other fuel. This method constitutes an evaluation in principle of the waste's capability of self-preservation combustion. But scientific requirements demands specific values for all the factors taking part in a physical-chemical change. Therefore, we apply the method suggested by K.W. Shin according to which we can estimate the calorific capability of waste under the conditions of real combustion, with the assistance of an adiabatic calorimeter (3). This method has been applied in the Federal Republic of Germany since 1965, and a nomogram was formed with its help.

The results for September - December 1987 are shown below (Table 1) and are with agreement with previous analysis in Greece - Rhodos.

Table 1

Organic Substances	% w/w	32,0 -	48,0
Paper	% w/w	11,5 -	16,0
Plastics	% w/w	10,0 -	13,0
Rubber, Wood, Leather, Cloth	% w/w	1,ε -	8,4
Glass	% w/w	10,0 -	22,0
Metal	% w/w	8,5 -	15,0
Inert	% w/w	0,7 -	2,0
Humidity	% w/w	23,0 -	34,0
Ash	% w/w	32,0 -	45,0
Maximum Calorific Value	Kcal/kg	2760 -	4600
Minimum Calorific Value	Kcal/kg	500 -	1275
Total Carbon	% w/w	32,28-	49,59
Total Nitrogen	% w/w	1,40-	2,96 (dry sample)
Ratio C/N		11,22-	27,71

These results allow us to think the exploitation of the wastes, either by combustion or composting them (4).

References

Papachristou, E., Rekkas, A.S. and Margellos, A.G. (1987). Environment Protection from Urban Wastes Deposition and City planning Location of Utilisation Installations, IVth International Conference, Technological, Energetic and Economical Environmental Problems, Vol. 1, 32-37.

Iurzolla, E. (1983). Problematiche ambientali ed energetiche nellincenerimento dei rifiuti, LXX-N10, 929, Roma, Maggio.

Standard Test Method for Heat of Combustion of Solid Fuel by the Adiabatic Bomb Calorimeter. ANSI/ASTM D 2015-66.

Waste Management in Developing Countries. (1986). (Ed. B. Thome - Kozmiensky, J.K.). Vol. 1, E F-Verlag für Energie - und Umwelttechnik GmbH.

AN EXAMPLE OF SEWAGE SLUDGE UTILIZATION IN AGRICULTURE AT THE MUNICIPAL LEVEL IN BELGIUM.

P.O.SCOKART[1], M.P.WALBRECQ[2] and J.P.BAS[2]

[1] Institute for Chemical Research, Tervuren, Belgium
[2] Intersud, Thuin, Belgium

Introduction

Sewage treatment has been largely developed in Belgium during the 15 past years. In Wallonia, the south part of Belgium, approximately 150 sewage plants are operating with a wide range of capacities (1). They are managed by association of municipalities which cover a well defined region.

Elimination of the sludges is an important environmental problem. The association of municipalities responsible for the south of Hainaut (INTERSUD) has contributed successfully to resolve it by promoting their valorization in agriculture. The success of this action is due to a favorable situation in rural areas with no highly polluted sewage sludge, a good information of the users, a permanent quality control and experimental research about the value of the sludges.

The treatment plants and the managing authority

Figure 1 gives the localization of sewage plants. Very great towns do not exist in this region so that small treatment plants have been installed in the vicinity of each city. The first ones were started ten years ago, the others have progressively followed.

There exist no important industry giving high risk residues in this region. The public authority have only to treat the sewage waters of small cities so that controlling the

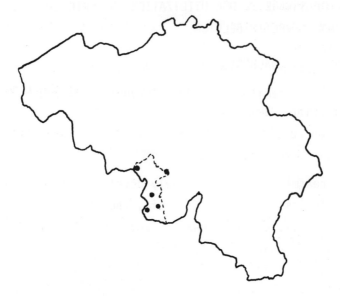

Fig.1   Localization of the sewage treatment plants.

sewage treatment process is not very difficult and the quality of residual sludges is suitable for agricultural use.

Intersud is the managing authority. It is an association of municipality for promoting the economic activity of the region, the resources of which are coming essentially from the Walloon region authorities.

Users information

A structure has been installed to make the information circulate between the farmers and the administrators of treatment plants. It was sponsored by the regional public authorities. It works in several ways :
- A booklet was diffused amongst the farmers to inform them on the sewage treatment process, sludge composition and expected agricultural results. It gives also information on reglementation and good agricultural practices when using

sludge in agriculture.
- Exhibition and lectures on the same topics are organized for the farmers.
- Personal contacts of the plant managers with the farmers located near the treatment plant contribute to **their decision** to start using sludges for their crops.
- Utilization conditions are tested and field trials are organized for demonstrations.

Quality control

Before spreading, sludges and soils are always analyzed for their nutrients and metals content. Two campaigns of sampling are generally performed in a year, corresponding to the **spreading period in March-April and in September-October.** Nutrients and metals are determined in soils by classical extraction methods (2). Dry solids content, organic matter, nutrients (N, P, K, Ca, Mg, Na) and metals (Zn, Cu, Cd, Pb, Ni, Mn, Cr, Fe, B, Mo, Co, Al, Sb) are determined in sludge. The samples are oven-dried for dry solids determination and calcinated for organic matter determination. Nitrogen is determined by the Kjeldahl method, the other elements are determined by ICP spectrometry after mineralization in a HNO_3-HCl (1/3) mixture. The results are given in table 1 and 2.

The composition of the sludges reveals sufficient quantities of organic matter, nitrogen and phosphorus to be valorized in agriculture. Calcium associated with the organic matter contributes also to improve the value of the sludge as a soil amendment. The quantity of potash is too small to be considered by the farmer.

The most important problem for the valorization of sludges in agriculture is their heavy metal content. As shown in table 2, the metal content of the sludges is below the guide-

Table 1 : Nutrient content of sewage sludges (% D.S.)

	Minimum	Maximum	Mean
Dry solids	1.3%	5 %	3.5 %
Nitrogen	2	12	4.6
P_2O_5	1.4	6.9	4
K_2O	0.2	1	0.5
CaO	1.2	6	3.5
MgO	0.7	0.95	0.8

Table 2 : Heavy metal content of sewage sludge (mg/kg D.S.) and guidelines values

	Minimum	Maximum	Mean	Guidelines Belgium	Guidelines EEC
Cu	100	300	160	500	1.750
Pb	50	400	180	300	1.200
Cd	7	15	10	10	40
Zn	220	2.400	1.600	2.000	4.000
Ni	13	75	33	100	400
Cr	40	130	75	500	-

lines proposed by the EEC and the Belgian Ministry of Agriculture. If the input on the soil is considered for a typical 50 m³/ha application (table 3), problems may perhaps appear for high content of Zn and Cd after a long term application on the same soil. As the farmers integrated the sludge dressing into the rotation, the delay is increased to 30 years. **By changing the fields and/or the farmer, it may also be in**creased by a factor of 2 to 3. It may thus be concluded that, due to their little content of heavy metals and a good rotation of the fields, the use of sludge in agriculture gives no problems. A control is regularly performed by analysing the **soil for their heavy metal content and until now, no** variation is observed.

Table 3 : Heavy metal input (kg/ha/cm) for a typical 50 m³/ha yearly application

	input	guidelines
Cu	0.28	12
Pb	0.32	15
Cd	0.02	0.15
Zn	2.8	30
Ni	0.06	3
Cr	0.13	-

Economic aspects

The elimination of sludges represents generally a great problem for the managers of the sewage treatment plants. Their valorization in agriculture is thus a good solution with little investment. In collaboration with the public authorities, they have to support the costs of the information campaigns and the analysis and control. They must also employ persons having good contacts with the farmers to assure the success of the operation.

The farmer himself has to support the cost of transport and spreading for the sludge. As the farmer is already equipped for the spreading of farm manures and the same equipment is used, the costs are calculated in terms of marginal costs. For a typical application of 50 m³/ha of a 3.5 % dry matter sludge, the farmer receive 90 kgs nitrogen, 70 kgs P_2O_5, 60 kg CaO and 10 kgs MgO. Field trials have shown that the efficiency of nitrogen is slow (from 25 to 50 % the first year), so that, as compared with mineral fertilizer, the fertilizing values of the sludges may be estimated between 3500 and 5000 BF/ha. This can cover the cost of transport and give a positive return if the fields are not located too far from the plants and the access of the plants is easy for farm equipments.

The value of organic matter in sludge is difficult to quantify, but it is highly appreciated by the farmers and it contributes also to increase the value of the sludges.

The quantity of sludge must also be adapted to the area of the field plots. In this case, the elimination of 20 to 50 m^3 sludge two times a year is perfectly suited for a typical 50 m^3/ha application.

References

Barideau, L. and Delcarte, E. (1986). Sludge valorization in Wallonia, SCOPE symposium on Belgian Research on Agriculture and Environment (Ed. M.Verloo), Brussels, 147-155.

Hoenig, M. and Scokart, P. (1983). Applications de la spectrométrie d'émission dans un plasma à couplage inductif (ICP) à l'analyse minérale en chimie agricole, Rev.Agric. 36 (3), 1727-1736.

Session 15

ISWA 88 Closure